# Science and Daily Life in the Study of the Northern Lights, 1898–1928

# Science and Daily Life in the Study of the Northern Lights, 1898–1928

By

Mette Bunting

and Toril Synnøve Jenssen

Translated by Josephine Stenersen

Cambridge Scholars Publishing

Science and Daily Life in the Study of the Northern Lights, 1898–1928

By Mette Bunting and Toril Synnøve Jenssen
Translated by Josephine Stenersen

This book first published 2018. The present binding first published 2019.

Cambridge Scholars Publishing

Lady Stephenson Library, Newcastle upon Tyne, NE6 2PA, UK

British Library Cataloguing in Publication Data
A catalogue record for this book is available from the British Library



ISBN (10): 1-5275-2424-8
ISBN (13): 978-1-5275-2424-8

The English translation has been financially supported by NORLA

# TABLE OF CONTENTS

# PREFACE

# HISTORY DEVELOPS SLOWLY, JUST LIKE LIFE

The authors of this book met for the first time in autumn 2007, when attending a conference about the Northern Lights at Alta Museum. Before that we had not known about each other's existence. Neither did we know that for a long time we had both been playing with a similar idea concerning the historical events, to which we were both closely associated.

Unique material was available; a collection of letters from daily life on a North Norwegian mountaintop, which had been a bastion for Northern Lights research since 1898. The originator for all of this was Professor Kristian Birkeland (1867-1917). Birkeland had developed a theory about the Northern Lights based on laboratory experiments at the University of Kristiania at the end of the 19th century. To continue these research results he established the world's first Northern Lights' Observatory on the summit of Haldde Mountain in West Finnmark.

This steep mountain with its 900 metres above sea level is among the highest mountains surrounding the small village of Kåfjord. A house was built as an extension to the observatory in 1912, and was further extended in 1915 to house more residents. In that rather bizarre place, where snowstorms can range from the end of August, there were, at the most, four families of adults and children living in the research colony. Three of the children were born on the mountain. The letters illustrate an interesting picture of everyday life and scientific research mainly from a woman's perspective.

The events on Haldde occurred when our grandparents' were young. Mette Bunting is a granddaughter of Dagny and Ole Krogness, who during the years 1912-1918 lived on Haldde. This was directly above the village where Toril Jenssen's mother's family lived. Toril's grandfather, Henrik Jensen, who was a carpenter, took part in the building of the house on Haldde.

During the Northern Light's conference in Alta in 2007, Mette gave Alta Museum, on behalf of herself and her sister, Bente Ødegård, two armchairs, a chronometer, and a lovely dress. Both the chairs and the dress had been used on Haldde during the height of its heydays, and the chronometer was used there, and in Tromsø, in connection with scientific measuring done by Mette's grandfather.

Toril had worked with the letters for a while in connection with a manuscript for a documentary film, and held a lecture about events on Haldde. The room was full of interested listeners. "Write a book about it," they said.

We followed their advice and have worked together on the material for this book, producing this monograph through our shared effort. However, while working with the material we have also individually taken more responsibility for some themes, as shown in the author's list at the back of the book, as well as contributing material for the other chapters.

We would like to thank everyone who has given us their time and goodwill so that we could realise our book project. Dagny and Ole Andreas Krogness' family willingly shared with us the material they had collected. There were letters, memoirs, a diary, photographs and newspaper cuttings. The descendants of Olaf and Dagny Devik, who also lived on Haldde for some years, have likewise, shared a lot of what they had of letters and photographs. A special thank-you to Hilde Holbæk-Hanssen, Ole Andreas Krogness and Hennige Astrup and other family members for collecting this material from these families.

Both in Kåfjord and in Alta, we have experienced great willingness from people who had memories and tales about Haldde. We would especially like to thank, Åse Synnøve Oppegård, daughter of Sanna Pedersen, who had been a maid on Haldde, Rannveig Håkegård Pedersen, local historian in Kåfjord and Werner Daniloff, previously headmaster of Kåfjord School and leader of the Cultural Committee in the village. To be able to obtain the stories local residents in Kåfjord still remember, has been enriching and important. We are very grateful for the interviews with the sisters Mary and Borghild Lukkassen about their childhood memories from Haldde. Also the interview with Gudrun Dagenborg, for her memories from when she was a maid on Haldde, and the conversation with Synnøve Holbæk Hansen, who was born on Haldde. These women gave us irretrievable contributions at the end of their lives.

Many have contributed with photographs and text material for the book. We would like to especially thank Frithjof Heitmann, Anne Marie Sørensen, Bernt Øien Jenssen, Reidun Tøllefsen and Nina Levin, who gave us access to their late relatives' photo albums and literature collections. Thanks also to Hallvard Birkeland and Per Eliassen, for good help in searching for special photographs.

Discussions with professional people have been important, both in connection with Northern Lights' research and historical research. These are: Researcher Magnar Gullikstad Johnsen and Professor Asgeir Brekke at UIT the Arctic University of Norway, and Associate professor Truls Lynne at the Northern Light's Observatory in Tromsø, Professor Knut Kjeldstadli and Professor Alv Egeland at the University of Oslo. Many thanks to those who have used their time to talk with us, and have corresponded, via e-mail. Thanks to NFF, the Norwegian Non-Fiction Writers and Translators Association for our project grants for the Norwegian publication.

Many thanks to Alta Museum that have on a number of occasions, allowed us to use valuable archive material freely. A special thank-you to the Museum's conservator, Hans Christian Søborg, for good advice and useful cooperation. A big thank-you to the National Library, for the free use of illustrative material, and their letter collection, and for developing photos from the glass plates, which was a demanding process. Thanks also to the Norwegian Museum of Science and Technology in Oslo, Tromsø Museum, the Regional State Archive, and Perspektive Museum in Tromsø for their help, when searching for photographs and other illustrative material.

We would like to give a big thank-you to Orkana Academic Publishers, for their belief in this project from the beginning and for publishing the Norwegian version in 2014. In connection with the translation from Norwegian to English we would like to especially thank Josephine Stenersen, retired Associate Professor, who has translated the book, and Mark Stenersen, Graphic Designer and advisor in visual communication and development, at RESULT, UiT The Arctic University of Norway, who has done a great job preparing and organising the final English manuscript.

A special thank-you to Kjell Kolbeinsen and Trond Mohn for actively contributing financially, so that the publishing of this book could be a reality, together with the following, who we give our very grateful thanks:

**Rector Anne Husebekk**, UiT The Arctic University of Norway
**NORLA** – Norwegian Literature Abroad
**Alta Municipality**
**Dean Arild Hovland**, University of Southeast Norway

Finally, we would like to thank each of our family members and close friends, both in the north and south, who have been sparring partners, and have contributed with local knowledge. They have been interested and a good support to us throughout the whole process.

Skien and Tromsø, August 2017-08-17

Mette Bunting and Toril Synnøve Jenssen

# INTRODUCTION

# CULTURE'S GARRISON IN THE NORTH

This is what their house looked like while it was being built. A tunnel to the observatory had been dug out and was still lacking a roof. Archive: World Heritage Rock Art Centre - Alta Museum

*Dear Mother, Yes I am sitting on Haldde in brilliant sunshine. You can believe me it is beautiful here. I almost walked all the way up, but for the last steep part, and then I sat on the sledge, which was buckled to the horse. Have you sent double woollen blankets?*[1]

[1] Letter from Dagny to her mother-in-law 02.10.1912

It was a stormy day in October 1912 that Dagny Krogness travelled on a local ferry into Alta fjord. She was 25 years old, newly married and her husband, Ole Andreas Krogness, was her travelling companion. He had been appointed as the Director of the Northern Lights Observatory, which was placed in a deserted mountainous landscape where they would be staying for the next six years. The quotation above is from the beginning of a letter to Ole's mother, Christiane Augusta Krogness, from a frozen daughter-in-law.

About a hundred years later in the summer of 2008 we find ourselves on the ascent from Kåfjord towards the first craggy rocks, along a trickling brook in a birch wood and we climb further upwards and upwards. The Krogness family are to have a reunion. A whole flock of Dagny and Ole's family descendants are on their way to Haldde, together with us, the two authors. The last bit is hard going, as it is very steep. Down below we can see houses that have been rebuilt after the Germans had burnt almost everything in 1945. Some of them are empty during the winter, but in the summer holidays many of their owners, who otherwise spend the rest of the year in other places, fill many of them.

It is a lovely summer day. We are almost at the summit and see a large building of natural stone in shades of grey. Like a castle towering towards the heavens. What traces of past history can we find here?

Haldde is a Sami word and means holy mountain. The mountain looks like a pyramid in shape and can be seen clearly from most of the central places in Alta and the community of Bossekop, which is nine kilometres further out in Alta fjord. You can catch a glimpse from there of the observatory and house, as two small black clumps on the mountain's summit.

The house is like a fortress built of large stone blocks. A darker and a lighter part of the building inform us that there has been an extension added to the house. The two-floor building is dominating. There are lattice windows on both floors, with solid frames. Around the house there is not much room to move, as the mountainside falls steeply down the pathway that we came up. At the back of the house there is a short slope up to the actual observatory. We are lucky with the weather. The view is fantastic from there.

We walk into the building, which has housed so many people, adults and children, researchers, wives, servants and assistants. The house is divided into two large rooms now after the building was renovated, which took place in the 1980's. The rooms are empty.[2] Only the outside of the

[2] Restoration work was done in 1980, under the management of Alta History Society, Alta Museum, and the County Curator in Finnmark with financial help from the Norwegian cultural fund. (See Nielsen & Petterson 1993)

building is the same as it was when people lived here. The tunnel that joined the house, observatory and outhouse has fallen down, and there is a temporary wooden roof over the passageway.

The research colony's home and work place. Archive: World Heritage Rock Art Centre - Alta Museum

We come up to the observatory, and look out over the mighty mountainous landscape, with Alta in the distance towards the northeast, and further on over the land towards the east and south. The mountains lie below us, and we can see the fjord a long way down below. We peep into the outhouse and return to the main building. Here were the flats and the maids' rooms, the office, cellar and attic. On that small spot, large-scale research was carried out on the Northern Lights, wind and weather-forecasts, rearing of livestock, births, maintenance of instruments, making meals and leisure time; work and private life hand-in-hand. Within these solid stonewalls life was unfolded.

When Dagny saw her new home on the summit of the naked mountain, she thought, "[…] it is not likely to be very warm in this house, but I beg your pardon, when I got inside I was met by light painted walls and it looked quite cosy".[3]

The Northern Lights' research on Haldde can be regarded as the most significant event in North-Norwegian history, as well as being nationally important. However women's contributions, have to a larger degree been omitted in written accounts. That three children were born up on the mountain, for example, has not been mentioned very much. The research, which was carried out on Haldde Mountain, is rightfully placed in Norwegian geophysical science development. It was a branch of knowledge known and valued outside Norway's borders, "Haldde, the cultural outpost in the north, really became an important place in Norwegian geophysics for more than a decade",[4] Professor Alv Egeland writes in 1994. Olaf Devik called the Haldde-period, a time of pioneering in Norwegian physics and geophysics.[5]

From an historical perspective, Professor Kjeldstadli, writes:

> […] at the same time research is only one side of life, and research is restricted by everyday life and budgets given by employers, not to mention the researchers' own intelligence and perseverance. Research is also about finding ways forward with the ability, the time and the resources one has.[6]

In this book we focus our attention on the everyday events surrounding the research activities on this wind-swept outpost. We will present the daily activities, which was the "glue" that kept the existence on Haldde Mountain together; and a significant premise for the creation of their intellectual work. We want to shed light on the social organisation round scientific knowledge. Everyday life does not restrict itself just to the home; it also covers the world as seen from the home.[7] Therefore this book also contains stories about how life on Haldde expanded and was spread to other arenas, and into other contexts. Daily life was part of a larger scene that had far-reaching consequences.

In publications by, "*Alta Historielag*", and articles in journals as well as in chapters and paragraphs in specialist literature, the activities on Haldde are described. We have especially had great pleasure in reading, "*Nordlyspionerene*", about the Northern Lights' pioneers, written by, Alf Ragnar Nielssen and Arvid Petterson, published in 1987. We have also

---

[3] Dagny's diary from 1912.
[4] Egeland 1994:19
[5] Devik 1976
[6] Kjeldstadli 2007:6
[7] Gullestad 1989:175

made good use of Olaf Devik's self-biography called, "*Blant fiskere, forskere og andre folk*", from 1971, which tells us about fishermen, researchers and other people. While working with this material we have also been inspired by the book, "*Mine fire besteforeldre*", written by Knut Kjeldstadli in 2010, about his four grandparents.

The main material, which we rely on in this book, is based on a private letter collection. Dagny herself writes the majority of letters to her mother-in-law. Some letters are written by Ole, and a few are answers from Dagny's mother-in-law. She was clearly an important contact in the couple's special life situation in the north. Altogether there are eighty letters of this unique material, kept by Synnøve Holbæk-Hansen, Dagny and Ole's daughter. Synnøve was born on Haldde Mountain in 1915 and died in 2006. From her and the rest of the family, we have been given permission to make public the contents of the letters, which today can be found in the National Library's hand-written collection.

We also base our book on Dagny's diary, which she started on the very first day that she lived on Haldde. Her diary is in Alta Museum. Other documents include Dagny's memoirs, which she wrote for her grandchildren; and this covers the whole of her family history. She completed these notes in 1976. We have collected supplementary material from interviews with people, who lived on Haldde, at that time, and from public archives where we have found information about the research activity, both in the form of letters and other documents. To be able to come even closer to what actually happened in everyday life on the mountain, we visited and discussed this with academic researchers working with physics and history.

Many of the incidents we write about can be visualised in the unique photo material, which we have been given access to. We have used some illustrations from a collection belonging to Alta Museum as well as from other archives. A large private photo collection from family members' photo albums and original glass photographic plates, which were donated to the National Library of Norway, have been free for us to use. We have also obtained photographs from descendants of people we mention.

Some of the personality of the people we present becomes clearer through the letters and photographs. We can put ourselves in their position and see how they participate in events that change over time, as new decisions are made. These activities move forward and into new forms. We try and present a coherent story about how life was experienced by those who were researching the Northern Lights, under extreme conditions in the Arctic. We follow Dagny and her experiences through a number of phases while living there. She is our main informant. What was important for her and those around her during these years?

As we write questions pop up, which have relevance to the closeness each of us has to the incidents that occurred. How well do we know the people, who participated in these occurrences? Can we place ourselves in their thoughts and feelings? This relates to research's ethical questions about how to make people and sources anonymous, how to present events and attitudes that can be related to cultural background and class affiliation. It is also how we approach descendants of those we write about and who are alive today. Our own cultural background is most likely to influence how we interpret our material. The impressions we have of people and social relationships that come to mind from our own knowledge, experiences, research and our own life, will influence our interpretations of events and people. The cooperative work, for this book, has also resulted in us taking a journey into ourselves.

The questions we ask are: What was it like for these people to live on Haldde, and what were the everyday activities based on? What did they achieve? Where did they go to when they left Haldde?

In trying to understand the events as being a meeting between different cultures from the north and south of Norway, we show how the local people are portrayed in some of Dagny's accounts. We have chosen a chronological approach, so that we can participate in life on Haldde as the years go by. We let Dagny, Ole and other participants tell about events using their actual words in the quotations, letters and other documents. In this chronology we show everyday life's diverse activities, births, hospitality and social occasions, loneliness, travel, scientific research work and the world around them. When Dagny and Ole touch upon important events, people and tendencies in their lifetime, we give supplementary information.

## Theoretical background and everyday life

A society that does not know its past will suffer from a collective loss of memory. Such a society does not know what they are, where they come from and where they are going, claims Knut Kjeldstadli.[8]

We have material about historical events in the Arctic zone. With a background in education and sociology our contribution takes us a step into an unknown terrain. The book project is interdisciplinary. We are closely connected to this material, as we have heard the stories from our

---

[8] Kjeldstadli 2007:23

grandparents about Kåfjord and Haldde Mountain. It is this that has made us inquisitive and, at the same time, aware of treating the material we use with respect, both with regard to the actual subject area and the people we present.

The letters document a segment of activities, as interpreted by the letter-writers themselves. Through the letters we come closer to the life of the people who wrote them. In these letters they tell about their experiences, adventures and thoughts. They place themselves, language-wise, according to their own cultural context, both in close relationships and distant social structures. They hold firmly onto the real moments in a written form, and are therefore able to deliver their thoughts to posterity. Their assignment on Haldde was special, and we can imagine that they believed that their stories would be of interest to future generations. At the same time, letters are always addressed to an addressee, so they are also important testimonies of relationships and networks in a specific contemporary culture.

We will look at the material from an everyday perspective. The choice of perspective is based on our interpretation of the contents of the letters, which is our empirical material. Our approach can be called a, "double hermeneutic", as we will interpret the interpretations that the letters convey.[9]

Even though the everyday-life perspective mainly acts as a background for the letter material, our own theoretical understanding may appear quite unintentionally in the choices we make, when using the actual texts. We do not want to just describe what the letter-writer describes, but to achieve a deeper understanding of these actual events. The material documents some of the things these people have thought and experienced, as well as some actions that they want to tell the recipient. Therefore these letters can be regarded as prime sources. Among the sources we have at our disposal, it is these prime sources, which are closest in time to what actually happened.[10] Obviously not everything is to be found in the letters. We know very little about what has been left out.

At times we have raised questions about things that might have happened, but for which we have no documented material. What have they speculated about? What did they talk about? What actually happened? We wonder. The letters, diary and other sources are remains of situations they experienced. They are documents that show just a part of the past reality in which they lived. When we read the letters or the diary notes, we only manage to get part of the situations that have been included. We can draw

[9] Giddens 1993

[10] Kjeldstadli 2007:177

our own conclusions about a particular person, or from a more extensive situation that resulted in the actual source material. From small pieces of information, it is possible to piece together a larger overall picture.[11] We reconstruct the past in this book, based on our source material and our interpretation of it.

Dagny wrote a diary, especially at the start of her stay on Haldde. Archive: World Heritage Rock Art Centre - Alta Museum

As we also try to interpret what is not said, we come closer to what sociologists call, triple hermeneutic.[12] It means that we interpret with a critical eye and search for the underlying cause, hidden interests and driving force, not just in the people we write about but also in ourselves. That our grandparents are the actual figures in the events, is a challenge that demands reflective consideration as to credibility in our interpretation of the material. In the interpretative sciences, the relationship between proximity and distance to the material is discussed, especially related to field work studies, where the researcher delves into a known local community, talks with people, and writes about the culture they are analysing. It has been maintained that an, "inside perspective", blocks new knowledge, because the researcher is too close, and knows the study-field so well that she becomes oblivious to it. One does not see what is obvious.

[11] Kjeldstadli 2007:229
[12] Alvesson & Sköldberg 1994

Closeness and distance to a field of study is important, but both positions offer advantages and disadvantages.[13] Even so the researchers' own personal assumptions will influence what material one can get access to, and what problems one is likely to meet when continuing to work with it.[14] This discussion is about studying one's own society versus an unfamiliar one.

This book contains grandparents and a rural community, which individually we know very well. The events happened a long time ago, and most of the people we mention have died. History has moved on and distanced these people from us. The distance of time means that even closeness includes distance. There is clearly not so much depiction of life portrayed a hundred years ago. Therefore, close proximity can, to a lesser degree, make us oblivious to the well-known situations, and block out new information, despite both of us having our own personal ballast with us in the way we approach these situations. We look at the past using the present day's cultural assessments as a background, and do our best when using the interpretative repertoire that we know. At the same time we try to balance a critical investigative perspective while taking ethical care of the people we present.

Using private letters as our main material is not without its problems. What right do we have to make public private communications from real people, by quoting from letters that have been written confidentially to a specific addressee? The letters are meant for just that person, not for the public. At the same time we know that the everyday descriptions in these letters are more of a report than intimate confidentialities. Most of the letters are about everyday life, which was also a working life for a housewife. The young Dagny's position, as wife to a researcher, was to take responsibility for organizing the household and home. It was virtually her specific task, as she was part of her husband's official position. That she kept the letters, when her mother-in-law died, can indicate that she thought they were important for posterity

What do we understand by the expression, "everyday life"? Everyday life is daily-life, that which quickly becomes a routine, and which we do not ask many questions about as it is taken for granted; it is "the unheeded".[15] In everyday life we maintain our existence, and recreate ourselves from day to day, while we sleep and while we are awake. Everyday tasks keep existence together, but it happens in a manner that does not demand so much attention, even if it is of great importance. In

[13] Rosaldo 1989
[14] Paulgaard 1997:75
[15] Hviid Jacobsen & Kristainsen 2005

everyday life we also find extremes in our existence, such as in birth and death. Everyday life is all-embracing.

Danish, Troels-Lund's 14 volumes, "*Dagligt Liv i Norden i det sekstende Aarhundrede*" is about daily life in the Nordic region in the 17th Century, and covers a wide range of titles about housing, clothes, food, parties, birth, engagements, weddings, marriage, morals, phases of life and life's termination as well as cultural history, countries and people.[16] Almost everything about daily life is covered in the large collective works of Troels-Lund. Eilert Sundt, Norway's first sociologist was also engaged in the diversity of the lives people lived, during their everyday. He also studied arts and crafts, morality and cleanliness in Norway in the 19th Century.[17]

The Norwegian anthropologist, Marianne Gullestad, emphasizes two main dimensions concerning everyday life. One is practical organisation of main tasks and activities. We find quite a lot about this in the letters. The other dimension is "lifeworld", which is about people's experiences. Lifeworld can be understood as the sum of that which is obvious and inevitable, and therefore covers the background of people's total existence, also about things that happen outside the home. In people's lifeworld non-specific knowledge that has filtered into the subconscious can be found, and included in everyday actions. This knowledge is justly significant, and is expressed in communication and linguistic formulation, and is associated with culturally belonging.[18] Home is the place where a person's different roles and participation in different environmental activities, is united. Therefore, everyday life scenarios from the home world constitute a basis for analysis and understanding reality, as well as an entrance into an understanding of society. In this dimension we find a dynamic in the letters, a kind of transgression to something new and until now inexperienced, processes in which Dagny participates.

There are many theoretical approaches to the investigation of everyday life, but there are principally two main approaches; a social-philosophic and an empirical-sociological one. The first is directed towards a structural level and makes use of abstract categorizing of the everyday. The other is mainly directed towards the specific, reality near, otherwise an everyday life containing "myriads of small things" that individuals do.[19] The connection between these two main approaches is exciting. It is about

---

[16] Troels-Lund 1914

[17] Stenseth 2000

[18] Gullestad 1989:25. Life-world concept comes from Hussel's phenomenological philosophy

[19] Maffesoli 1993: 7, ref. in Hvid Jacobsen & Kristiansen 2005:22

macro and micro levels in society, and how these levels stand in an interrelationship with each other. In recent understanding of everyday life, it is important to avoid a polarisation of a one-sided sociological macro study or one that is a narrow-minded micro-sided study.[20] Wanting to describe society "from below", implies a risk of being drowned in empirical data, which means just describing instead of analysing.[21]

We want to give a detailed description from everyday life in extreme surroundings, but we also want to take a look at daily life events from the point of view of other contexts, so that we can take an analytic look at the larger relationship in which the everyday situations are an integral part.

In the Haldde material the relationship between the individual and overall structure, is open. This is the actual heart of the matter and is documented in each of the individual letters, the majority of which are based on an arena that is a woman's domain. Our perspective will especially be from a woman's point of view, with opinions from experiences of everyday life. In a strange way these experiences are connected to the "extra-local", level of society and in that way, these experiences gain a retrospective strength in the organization of new everyday situations. The "extra-local" level is a social sphere, characterized by specialized institutions, formal organisations, expert knowledge and abstract work, also called society's dominant system.[22] In this case it constitutes the geophysical research environment, and the central authorities constitute the most influential dominance system, in addition to cultural values in the individual's lifeworld.

Maybe the theory of everyday life can be a meaningful background in order to see some contexts that otherwise would not appear. Even in the most extreme living conditions and special situations there is an everyday life, which relates to an extra-local society, which both affects downwards and becomes affected from below.

Our ambition is to present the empirical material in such a way that it will contribute to understanding everyday-life, where structures take root, and either arise and are cultivated through the personal life of the subjects own development and initiative, or they come from overall systems, such as cultural expectations, or clear commands and limitations, which merge together with everyday experiences.

---

[20] Hvid Jacobsen & Kristiansen 2005:30
[21] Gullestad 1989:37
[22] Aakvaag 2008:199ff.

# CHAPTER ONE

# A DRAMATIC ENTRÉ

Dagny Vegerd Guldberg and Ole Andreas Krogness out rowing when they were engaged. Archive: National Library, Photo collection.

*After a lovely trip along the Norwegian coast up to Hammerfest, continuing to Alta, with that terrible boat, "Nor", we arrived in Kåfjord after one and a half day's journey.*[23]

*We have reported the boat, as it was dangerously loaded and filthy. Everyone in the whole of Alta has complained.*[24]

The 26th June 1912, after the couple had been engaged for four years, Dagny Vegerd Guldberg and Ole Andreas Krogness, celebrated their wedding day in Kristiania, (Oslo). At last they can be married as Ole has completed his final examinations as a Master of Science. Both of them are

[23] Dagny's diary from 1912

[24] Memoirs for grandchildren written by Dagny in 1976

in their twenties, Ole 26 and Dagny 23. Towards the end of her life Dagny sums up the wedding in three sentences, "When he had completed his examinations we got married. There were 70 guests at the wedding. It was my aunt Julie that held it for us at the Studenterhjemmets Hotel."[25] It was not very long before Ole, on 1st July, was appointed Director of the observatory on Haldde in Finnmark.[26] Immediately after receiving this appointment, the newly married couple left for Germany on a two-month scholarship. Ole was going to do some research on terrestrial magnetism, his special field of study, and also to buy instruments for the observatory.

After their stay abroad they prepare their journey north, as Ole is to begin his new job 1st October. Ole has a lot to do with all the preparation of packing important equipment for the observatory, as well as the normal household necessities. The world's first Northern Lights' observatory, is placed on a mountain, 900 metres above sea level and is to be their home and workplace, with Dagny's maid and Ole's assistants. The actual building is a long way off the beaten track and there are no neighbours. From Kåfjord it took about three to four hours to climb up to the house. Their luggage is to be transported by horse and sledge, but they have to walk most of the way. It is only during the winter months, September to April that the house can be lived in, as there is no water nearby and they depend on using melted snow.

The couple do not know anyone in Kåfjord and they have no family in Finnmark. Their mothers, who are widows, live in Kristiania. The contrast from city-life that they are used to in the capital city will be rather different for the couple, who are emphatically social and closely attached to their family; according to the description given of them by their descendants.

In September, Ole and Dagny leave Kristiania and begin their journey. It is the year that the whole of Norway cheered over Amundsen and his men, who had reached the South Pole. Little does Ole Andreas know, as he is busy packing for the move, that the work he is going to begin would later bring him and Amundsen together, in connection with work in the Arctic region.

Other events in the world that left their mark on 1912, were that China became a Republic, and that the USA's president, Theodore Roosevelt, was shot, but perhaps what is remembered most of all is that the Titanic was shipwrecked 14th April. The world's largest and safest passenger ship of that time sank and resulted in 1,350 people losing their lives.

---

[25] Memoirs for grandchildren, written by Dagny, 1976

[26] Lindeman 1924:41

It is a long journey the young couple set out on. The complete journey is 1,800 kilometres, about 1,100 British miles, first from Kristiania to Trondheim by train, and from Trondheim to Hammerfest by boat.

The young Dagny and Ole Andreas. Archive: The National Library of Norway - Photo collection.

In 1912 Hammerfest is the hub of the shipping trade for the whole of Norway, even the Hamburg route[27] docks there. Here the couple must change boats and travel on a smaller local ferry, which transports people and goods into Alta fjord and to Kåfjord, where the couple are to get off. Usually, a newer boat called "Alten" was used for that journey, but the reserve local boat, "Nor"[28] is put into service, and is a lot smaller and older than the one it replaced.

Dagny has previously packed the most important and private belongings into three large suitcases. These contain clothes for all kinds of weather, pictures and other personal belongings. They also have some furniture with them that they thought they might need in their house on Haldde, plus boxes with other equipment for their home. Ole has

[27] From 1870 there was a ferry route from Hamburg to Vadsø (Nielsen 199: 378)
[28] Hansen 1999:7

instruments, for measuring the earth's magnetism with him, and camera equipment to take photos of the Northern Lights. He has also taken his scientific work, a great number of copies of magnetic measurements from his stay in Germany. He plans to do some more research on this work, and complete it. All of this is most carefully packed into large boxes, which are difficult to handle. Ole has an agreement with the steamship company to reload them onto the next boat when they reach Hammerfest.

In Hammerfest, problems arise, when they are about to reload their luggage onto the local ferry, "Nor". On checking that their luggage is in place, and that everything is as it should be, Ole finds that it is nowhere near ready. Dagny writes, "The reason why Ole had to work so hard to get the instrument boxes in place, was because the First Mate had been drunk since the middle of the day, and he had promised to put our things safely on-board."[29] Nevertheless, Ole manages to organize things so that the luggage is to be taken on board before the boat leaves that evening. The couple leave the boat, and set out for the centre of Hammerfest.

When they return to "Nor", in the evening, the situation is no better. They are not only met by a drunken First Mate, but, "[…] when we got there that evening both the First Mate and the machinist were so drunk that it was impossible to use them; the Captain had to stand on watch all night."[30] In some way or another the luggage is loaded aboard, the letters do not say anything about how. Dagny does not like the situation, and is disgusted that the ship's crew is to such a degree influenced by alcohol. Luckily the Captain is sober, and is able to stay awake through the whole night's sailing. When Dagny and Ole go to bed in their cabin, we can imagine that they are a little anxious, and that they long to get to Kåfjord, which according to their plans will take twenty-four hours.

What influenced Dagny to choose such a life with Ole? To take such a long journey with him, and to contribute to building up something, almost from nothing, so far north? During the short time they have been married, they have not owned their own house, and now they are on their way to create their first home, and that is on the top of a mountain, isolated from the surrounding world. In Dagny's family, such long and challenging journeys were not uncommon. Three of her brothers have already emigrated to America, and a fourth, who wants to live in a warmer climate, travels later to the Galápagos Islands.

Dagny's family were noted for their interest in travel and their pioneering spirit, and it is something their children have inherited from them. Before Dagny was born, the family were stationed on Madagascar.

---

[29] Letter from Dagny to her mother-in-law, 19.10.1912

[30] Letter from Dagny to her mother-in-law, 19.10.1912

Dagny's father, Carl Johan Guldberg, was converted when he was a young man, and wanted to work as a missionary on that island. He had to give up this work as he had emphysema, and his breathing problem made it difficult to preach. In addition, the rules of the Norwegian Missionary Society did not allow someone with such an illness to be in that climate. They believed it would be detrimental to his health. Carl Johan Guldberg was resolute, and instead, educated himself as a doctor, because the association for that profession was not subject to such health requirements. In the same way as Dagny and Ole, he and Dagny's mother, Louise Henriette, travelled to distant areas of the world as a newly married couple.

There was little doubt that Guldberg's work in Madagascar had been important for many people. Guldberg's reports to the Missionary Society[31] described all their duties and enterprises. He worked as a doctor, surgeon and dentist, and in that way saved many lives. The couple were idealists and paid for their own journey. The intention was that they would use the profit from the doctor's practice to buy medicine, and to help to build the mission-station. Even so, it appears that Dagny's father often worked without being paid, as the inhabitants were extremely poor.

Having textbooks in Malagasy, the Madagascan language was quite an issue for Carl Johan Guldberg. He believed that it was important to educate Madagascan health-workers. Together with a colleague, Dr Borchgrevink, he wrote and translated among other things, medical books for doctors and midwives, as well as books for pharmacists, surgeons and botany for doctors. Dagny explains that first of all her father worked with Dr Borchgrevink, but later when Dr Borchgrevink travelled back home, her father was left alone with the Negros.[32]

Dagny writes that her mother was good at telling stories, "My mother used to tell us dramatic stories about the war and escapes and many other strange events, during the long winter evenings throughout our childhood."[33] Dagny's mother had done more than observe things that were happening around her. She had altogether six births in ten years, and her main task and responsibility was for the children. The most difficult thing that happened to her was the loss of a little girl, Henriette Amalie.[34]

The family decides after ten years to return home. Dagny's father writes in his letter to the Governing Missionary Board that the reason for

---

[31] Carl Johan Guldberg's letter to the General Council in the Norwegian Mission Society, 1876-1886

[32] Memoirs for grandchildren written by Dagny in 1976

[33] Memoirs for grandchildren written by Dagny in 1976

[34] She died when she was three months old

them leaving the island, in addition to his own poor health, was his worry over his sons' upbringing and development; "[...] that my children are somewhat behind in Norwegian than others, [...] and my children are so engrossed in the Malagasy language and Madagascan morals that it will be difficult to make anything out of them later."[35] Guldberg believed that the boys' upbringing and education would only be good enough in their native country, and if they stayed any longer on Madagascar they would be unsuitable for a career in their homeland.

On their return to Norway their father said no to a job as chief physician at Louisenberg Hospital. He clearly preferred to travel to Dovre and be a district doctor. His reason was simple: According to him people in the towns had grey-black lungs, whereas the lungs of those who lived in the country were pink. Dagny was one and a half years old when they moved to Dovre.

The quay in Kåfjord, where the Krogness couple landed after an extremely difficult journey. Archive: World Heritage Rock Art Centre - Alta Museum

Security was in many ways what characterized Dagny Guldberg's adolescence. She writes in her memoirs to her grandchildren that as a child, she had a free and good life. Her older brothers, who she calls in her memoirs, "some real children of nature", included her in their games. The woods and fields were their playground, and animals and plants were their playfellows. A tame crow, a butterfly collection, and a family of mice were invited into their home.

Dagny's mother was in charge of a large household on Dovre. In addition to five boys and four girls, they had two maids, a governess, a cook and a Nanny. A Nanny, hired by their mother, was expected to be musical. "She believed that music was necessary for a good upbringing", Dagny writes in her memoirs. Nevertheless, she commented in the same

[35] Carl Johan Guldberg's letter to the General Council in the Norwegian Mission Society, 25.02.1885

notes that she experienced a level-headed upbringing. Their life was at times somewhat frugal, because their income varied, despite Guldberg's good job. Dagny tells us that even when they lived on Dovre her father was reluctant about his patients paying him, because many of them had very little money. Therefore, as far as she remembers, they were constantly in need of money. Some of the patients paid with products, something that meant that their diet could vary according to what the patient used for payment, which at times could be rather monotonous for the large household. Dagny writes about when her father was given a sack of sago as payment, something that meant that they had to consume a lot of sago soup and sago porridge. Even so, one can take it that Dagny's life as she grew up, the daughter of a district doctor, was economically easier than for many others.

We are not quite sure when Dagny first moved to Kristiania, but suppose it must have been while she was a teenager. She writes in her memoirs to her grandchildren that she started school in the capital city, and that things did not go so well. Dagny, like her brothers, had been educated at home, she had a governess, and her brothers had tutors. "In town we were thirty girls in the class, and they used a different method than the one I was used to."[36] It was a big change from a quiet life on Dovre with her brothers and sisters around her, to a school's miniature society where she did not know the ground rules. Dagny calls it, a "horrible" time. She went from experiencing that she was clever in her school at homes, to no longer doing so well.

As an adult, 21 years of age, Dagny moves from her home for good, to work in the capital. Dagny's older brothers already live there and are studying at the university. They found a place to live with their aunt, their father's sister, Mathilde Sofie Waage. She is married to Professor Peter Waage, a well-known professor of chemistry, who works with her father's brother, and develops what was later known as, "Guldberg and Waage's Law", the law of equilibrium.[37] Because Dagny was a woman, she was not given a chance to study, but was to live and work as a receptionist for another aunt, Julie Sand, her mother's sister and Principal of a residential home for students in Kristiania. Her wage was food and board and 15 kroner (ca. £0,81 in 1910[38]) a month. Dagny found this was impossible to live on. Just having suitable clothes on that wage was difficult, she felt. To

[36] Memoirs for grandchildren written by Dagny in 1976

[37] Pedersen (2014)

[38] Reference: National Bureau of Statistics, http://www.ssb.no/kpi (ca. 1910-17), http://inflation.stephenmorley.org/ Historical UK inflation rates and calculator (ca. 1910-17)

be economical with her wage she kept her dresses for as long as she could by using an apron. In that way, a dress did not get worn out or dirty too quickly.

If we work out what Dagny's wage is according to today's value, she earned 839 kroner a month, on top of her food and board.[39] It was at that time a normal wage for women, for the type of work she was doing. As a comparison to Dagny's 15 kroner (ca. £0,81[38]), a sawmill worker earned 96 kroner (ca. £5,30[3838]) a month, but an ordinary seaman earned 43 kroner (ca. £2,38[38]). Women were not expected to earn their own living to keep a family, and therefore earned less than men. In other kinds of women's work where food and board was included, such as being a maid, the wage was 18 kroner (ca. £0,97[38]) a month. At that time one used more of one's earnings on housekeeping and food, than we do today, a whole 60 per cent.[40]

Dagny met Ole at the students' residence. He has his major subject in mathematics and works as an assistant for Professor Birkeland. Originally, Ole was from Trondheim, born 23rd May 1886, into a family of traders and governmental officials. Ole's mother, Christiane Augusta, is from the musical Lindemann family, and music was something Ole could not live without. He mastered many instruments, such as the piano, violin, viola, cello and clarinet, as well as being a competent singer. Some of his sheet music and instruments are in the luggage on board the ferry. The piano is to come later, as it is not possible to take it with them on the journey.

Ole's mother is a widow, for the second time, when Dagny meets her. Her mother-in-law has a daughter from her first marriage with the chemist, Jacob Johannes Carsten Hirsch, who died in 1881 after being married for six years. Three years later she married Ole's father, Ole Andreas Tangen Krogness, residential Chaplain in Our Lady's Church in Trondheim. Beside Ole they had a daughter, Anna Karithe. She becomes a good support for Dagny. When Ole was only one year old his father died.[41] Maybe that was the reason why there was such a close bond between Ole and his mother?

---

[39] National Bureau of Statistics 2013
[40] National Bureau of Statistics 2013
[41] Lindeman 1924:39

Ole's mother's surname was Lindemann before she got married. The family were known as being good musicians. Here Ole's family are playing together. Ole on the violin, Christiane Augusta (Ole's mother) on the piano and Anna is singing. Photo: Ole Andreas Krogness. Archive: National Library. Photo collection.

Dagny grew up with a number of strong and independent women as examples in her family. The closest is her mother, Henriette Louise. Despite having a large family to manage, Dagny says that her mother had a great capacity and engaged herself in work outside the home, when they were on Madagascar. "She started a children's home for orphaned children, mainly small girls, I think it was children who were unwanted and put out for the wild animals. They were in a choir that she started together with them, and they sang very well."[42]

Her mother's stories about these children, and the native women who became blind, due to the wretched working conditions they had when they made lace, was a story that Dagny took with her into her adult life. After Dagny's father arrived in Norway, he immediately began to long for his work in a warmer climate, but her mother soon put her foot down when he began to talk about a new call to the Congo. Enough is enough!

The other woman Dagny looked up to was her strong and enterprising aunt, Cathinka Guldberg, her father's sister. She was Norway's first deaconess. Dagny described her as, "a lovely person, strict but as if

[42] Memoirs for grandchildren written by Dagny in 1976

something holy shone out of her".[43] All of Cathinka's brothers had become academics, which was something quite remarkable at that time. These were honoured in a family portrait with Guldberg, the parish vicar in the middle, while Cathinka, although a woman with an education was left out of the family portrait, in the same way as her sisters.

Portrait of Dagny's grandfather and all his sons. He did not want his daughters, including Cathinka, in the photo. Archive: Private, Mette Bunting.

They are clever women, especially Dagny's aunt, who had managed to accomplish so much, without a man at her side. On the other hand, Dagny chose to be with her husband at his workplace. Dagny's children and grandchildren knew her to be an explicit and independent person. She uses this strength of character to support her husband and his work, and together they built a life around a research station on the top of a mountain. She got the chance to show both independence and strength in events that she met later in her life.

[43] Memoirs for grandchildren written by Dagny in 1976

She is already able to show this ability on their journey into Alta fjord, on the ferry "Nor". Dagny has probably discovered that she was pregnant before the journey, and is now almost three months on the way. She does not say anything about this, in her letters or diary. It is only in later letters, when she mentions the date of birth that we learn about her pregnancy.

Dagny rests on the boat. Archive: Norwegian Museum of Science & Technology

The last one-and-a-half days have been difficult, with a crew she experiences as being drunk, but things will become worse. In the middle of the night, at 2 a.m. when they are not very far from Kåfjord, the boat tilts and begins to take in water. The couple wake up, and for Dagny, who cannot swim, there must have been some hair-raising minutes. Luckily, it is the sober Captain who is steering the ferry, not the drunken First-mate. Dagny writes to her mother-in-law that if it had been the First-mate, who was steering, she was sure they would have sunk. Dagny is furious, and uses words such as "absolutely horrible" and "like a pigsty" in her letters to describe "Nor", and she really wants to report the whole crew, so that they could be sacked.

Luckily, the boat does not sink even though there is water on the lower deck. When they sail the last stretch into Kåfjord, the couple are ignorant about any possible damage, and how much of their luggage is under water, and what this means for their private things; Ole's instruments, photos and negatives of magnetic measurements.[44]

[44] These measurements were registered on photographic paper that had to be developed. Usually there was a paper for each day.

# CHAPTER TWO

# KÅFJORD'S INTERNATIONAL HISTORY

Kåfjord with the Copper Works and all the slag heaps after mining for copper. Archive: World Heritage Rock Art Centre - Alta Museum

*Copper had been found in Kåfjord some years previously and that was why there was a church there, and a closed down copper mine.*[45]

What kind of a place was it that Dagny came to? Maybe she knew that when the first Northern Lights were measured there, the village was a flourishing industrial community round Kåfjord Copper Works. With the help of English capital, the works started running in 1826. The ore was shipped out through Kåfjord and Alta fjord to an international market. In 1840, Kåfjord's Copper Works' was the absolute largest iron and copper

[45] Memoirs for grandchildren written by Dagny in 1976

works in the country, according to the number of employees.[46] Kåfjord stood out from other adjoining rural areas with its town-like character. The Copper Works' director, an English engineer, Steven Henry Thomas, began a number of social enterprises for the workers, such as schools, health services and a kindergarten. He also built a church and started a bakery, a library, a dance hall, and a theatre, as well as introducing cash payment for transactions.[47] Local theatre performances were reported in the newspaper, "*Tromsø Tidende*".[48] Steven Henry Thomas also had a large house in Kåfjord, to which he invited travelling guests.[49]

Guests arrive in their yacht. The Director's house is in the background. Photo: Olmar Egenæs Archive: World Heritage Rock Art Centre - Alta Museum

Here is a quotation from a report about the scientific expedition. *"La Recherche",* which gives a glimpse of the old Kåfjord, as well as the Director of the Copper Works' stately hospitality:

---

[46] Drivenes 1984:112

[47] Of the buildings that are left, the church is the only one still standing, after the Germans' burned the other buildings during World War II.

[48] The newspaper: "*Tromsø Tidene*" 13. February 1842

[49] Scheen 1957

> Some hours later we were able to sit down round the table in the Norwegian house and cast our eyes on a veritable Homeric roast beef. Round the table was the factory's Director, English by birth, a German engineer Ihle, who had just taken his final examination as a mining engineer in Freiburg, an Irish gentleman, Mr Thomas who was the leader of the company, a priest from Finnmark on visitation, a Sami priest, who the Swedish King had so kindly sent us, to be our guide, a Russian nobleman, who wanted to see Nordkapp, four Swedish and Norwegian tourists and ourselves, who in this cosmopolitan gathering represented our dear country, France.[50]

The rural community in Kåfjord, like the rest of the region, had long been composed culturally and linguistically of mixed cultures. Swedish, Finnish, Russian and English, as well as Norwegian and Sami languages were represented in that cultural conglomeration of a rural society.

The Sami people have their original settlement and roots here, according to more recent discoveries of rock carvings, right back to 8000 BC.[51] They became integrated with Norwegian and Finnish immigrants in the 17th century. Together with the rest of the Nordkalotten region, Alta was also part of a struggle for power between the three states Norway-Denmark, Sweden and Russia, each demanding their right to take land taxes, as no frontiers had been established. This battle had culminated in 1609 when the Swedish King Karl 1X began the building of a fort by Alta river, but was driven off by the Danish-Norwegian bailiff and his men. After that the Norwegian King, Christian IV, secured Norwegian hegemony, with the help of Norwegian lumberjacks, by building a royal fortress on Årøya, an island in Alta fjord, Simultaneously the Sami people were driven from pastureland for their reindeer.[52]

In the 19th century, the next large Norwegian immigration took place in connection with the running of the Copper Works. The constitutional law of 1814 at last began to make headway, with regard to Norway becoming a Norwegian State, and it stressed the significance of a Norwegian language. After 1850 the Norwegian authorities pronounced that they intended to strengthen the Norwegian population and impair the Sami and Finnish (Kven) society.[53]

---

[50] Xavier Marmier: Concerning the visit to Kåfjord in 1838, reproduced in Scheen 1957:17

[51] Helskog 1988:67-76

[52] Nielsen 1984:103

[53] Nielsen 1995:328-330

Copper is washed and shipped out through the fjord by boat. Archive: Observatory on Haldde's archive in the Regional Archives in Tromsø.

The history of the Copper Works goes way back to the discovery of copper in the mountains over Kåfjord. From the 17th century there had been explorations for ore in Kåfjord, something, which interested two English merchants, who did business in Hammerfest, John Rice Crowe and Henry Dick Woodfall. It is possible that they believed in, and had ambitions to start a mining industry up in the "Polar Night."[54] Crowe applied for permission to search for copper, and in 1825 he was given permission to search in Kåfjord. Unfortunately, the result of his discovery was too small to make it worthwhile for business connections in England, to invest their capital in a mining industry.

[54] The whole of this story is described in an article by Pedersen (2008); a local historian in Kåfjord.

The Copper Works' attracted visitors. Photo: Olmar Egenæs Archive: World Heritage Rock Art Centre - Alta Museum

The Copper Works with their leaders, including the Director, Otto Witt. Workers can be seen in the background. Photo: Olmar Egenæs. Archive: World Heritage Rock Art Centre - Alta Museum

Marit Alslaksdatter, from Dullan in Kåfjord, made a big and rich discovery of copper. While she was wandering on the mountain, she suddenly found a stone with a different colour than usual. She was searching for a goat that had not returned home when she found the stone half-hidden in the green moss. It was as if the mountain shone like gold towards her. She tied her headscarf to a nearby birch tree, so that she could find the place again, and took a couple of the loose shiny stones with her.

After a while, and in agreement with her husband, she showed her find to shopkeeper Klerck in Bossekop, who contacted Crowe in Hammerfest. On the 17th May 1887 Marit signed a document, in which she gave permission to Crowe to take over her eventual right to the copper ore. She received 900 kilograms of Russian rye flour as payment. The Marit-mine came to be the most important and safest ore reservoir throughout the Copper Work's history, which stretched from 1826 to 1909. Marit was given a pension from the Copper Works and is mentioned in the Church register as the person, "who discovered Kåfjord's Mine."[55]

When Crowe started searching for ore, there was only one family living in Kåfjord. It was the Copper Works that attracted a workforce from all the Scandinavian countries to Kåfjord. Many people, at that point of time, were extremely poor, as there was little work to be found. The population figures in Kåfjord rapidly arose to 3000 at the most.[56] The workforce at the Copper Works' was to a certain degree recruited from the local population, but also from surrounding North Norwegian rural areas, as well as some from further south. The reason for the latter was, because although the leaders of the mines wanted a permanent workforce, the local people in the area needed seasonal employment so that they could also fish and farm.

---

[55] Pedersen 2008:62

[56] From the population census in 1835, Nielsen 1995-69

## From Copper Mine to Northern Lights' Research

The Director's home called, "The House", was built in English style with a bowling alley in the front. Photo: Olmar Egenæs. Archive: World Heritage Rock Art Centre - Alta Museum

The Director of the Copper Works called, "King Thomas", was an ardent admirer of the Northern Lights. In 1837 he founded "*Videnskabsselskabet i Kåfjord*", a scientific company, and did some measuring of the Northern Lights. A forerunner to both King Thomas's engagement with the Northern Lights in Kåfjord and the observatory on Haldde was the French Duke, Ludvig Philip, who during a round trip in Northern Norway in 1795 was fascinated by the mystical green lights.

An educational awakening in the age of Enlightenment stimulated this interest and an admiration for knowledge and rationality, and an equal detest of faith and dogma. There were many myths and theories about the beautiful and unpredictable Northern Lights. Right up until today the Northern Lights have been connected to superstition and mystery.

The Northern Lights over Bossekop drawn by Bevalet, who was with the "*La Recherche*", expedition, sent out by the French King to the Alta area in 1837. Archive: World Heritage Rock Art Centre - Alta Museum

In the Nordic region it was a popular belief that it was linked to dead souls, as well as being used for warnings about the weather.[57] Many researchers of that period tried to explain this Northern Lights phenomenon.

When Ludvig Philip later became King of France, he equipped three expeditions to Northern Norway. A natural science researcher, Paul Gaimard, led the, "*Gaimard-expedition*" ("*La Recherche*"), and they were based in Alta during the winter 1838-1839.[58] A number of French scientists took part in the expedition together with some recruits from the Nordic countries. Among them was the well-known botanist, Lars Levi Læstadius, who later translated the Bible into the Sami language.[59] While they were in Alta they came in contact with Director Steven Henry Thomas, who reported to the Frenchmen that he had seen a Northern Light that had been so low that it looked as if it was between him and the mountains in Kåfjord.

The researchers took many measurements of the height of the Northern Lights, and the expedition's own artist, Bevalet drew a number of lifelike illustrations of the phenomenon, driven by the romantic spirit of the times;

[57] Brekke 1982:56
[58] Knutsen & Posti 2002
[59] Nissen 1941:168

with a fascination for the exotic, and the wild and untouched nature. As a result of that expedition, "*Voyages of Scandinavie*", a great work of 31 books was published, full of beautiful drawings and descriptions.[60] Thomas's own observations can be found in book 11.[61] This expedition created the basis for scientific research in Northern Norway, and placed Kåfjord on the map for scientists interested in doing research of the Northern Lights.[62]

Under the first International Polar Year in 1882, there was another expedition to Bossekop to study the Northern Lights. Professor Aksel Steen, from the University in Kristiania, led the expedition. A Polar station was established, to observe the Northern Lights, in Bossekop. The same year the researcher, Sophus Tromholt, from Denmark, also went to the Alta-Kautokeino region to study the Northern Lights. While he was there he tried to take photos of the Northern Lights from Kautokeino, but he was unable to capture the moving beam of lights that danced over the heavens. Instead he had to be satisfied with taking photos of instruments and situations when the Northern Lights were being measured, as well as taking a great number of photographs of people he met in the heart of the Sami region.

In the book, "*Nordlysets Straaler*", about the Northern Lights' rays, published in 1885, Tromholt describes a visit to Kåfjord. What he found was a decaying rural district, because the Copper Works' had come to a standstill due to a fall in the price of ore. Tromholt does not mention any registration of the Northern Lights in Kåfjord. Director Thomas had left the settlement a long time previously. Tromholt's written work was translated into a number of languages, and during the last years of his life he travelled round the world, holding lectures, where he praised the Alta region's "favourable conditions for observing the Northern Lights."[63] The German Martin Brendel was the first person in 1892 to successfully take a photograph of the Northern Lights seen from Bossekop, as the Germans had also sent an expedition to the north to fetch material for their Northern Lights' research.[64]

That Birkeland had searched for an appropriate mountain to observe the Northern Lights in the above-mentioned region was therefore not accidental. In 1896 he had developed a theory about the Northern Lights, based on laboratory experiments, at the University in Kristiania.

---

[60] A complete set of this work can be found at UiT Norway's Arctic University in Tromsø.

[61] Lottin, mfl. 1843-46, reference in Scheen 1957:19

[62] Nielsen 1995:159

[63] Brekke 1982:58-59

[64] Brateng 1980:68

Researchers on a boat trip in the fjord 1902-03. Photo: Olmar Egenæs Archive: World Heritage Rock Art Centre - Alta Museum

At that time researchers did not agree about what the Northern Lights were and how they were formed. Birkeland believed that he had the answer; all he required was resources for his research work so that he could convince others. In Birkeland's theory the Northern Lights consisted of negative charged electric particles from the surface of the sun. These are moving at great speed, come into the earth's magnetic field and are drawn towards the poles. When they hit the earth's atmosphere, the atmospheric gasses light the particles and the special moving light appears.[65] In the Alta region far north he wanted to confirm his theory through research in natural surroundings, to finally be able to distinguish what kind of being the Northern Lights actually were, and at the same time consolidate Norway's honour as a research nation. In so doing he joins other National heroic figures such as Fridtjof Nansen and later Roald Amundsen, who raised Norway's reputation with their distinguished expeditions in the Polar Regions.[66]

[65] Brekke & Egeland 1979:77
[66] Drivenes & Jølle 2004, kapittel 2

The observatory on Haldde during the building process. Archive: World Heritage Rock Art Centre - Alta Museum

Birkeland, sitting with his back against the wall, in Kåfjord 1902-03. Photo: Olmar Egenæs Archive: World Heritage Rock Art Centre - Alta Museum

This was not a simple matter for Birkeland. He was already an experienced researcher, but his theories were often too imaginative for the experts of that period, and many were rather reserved towards him as a scientist. Today, now that the closer vicinity of outer space has been chartered in detail with the help of rockets and satellites, Birkeland's work has become more actual and interesting. His theory about the Northern Lights has been accepted, and later carried on into researchers' dissertations.[67] At that time he was met first of all with opposition from the official authorities, who were responsible for giving grants. His plans were too airy, was the reason they gave. Finally, he was allotted 5.000 Norwegian Crowns to finance an expedition to the Nordic regions. The Government had at last understood that it was important that a Norwegian physicist should research the Northern Lights.[68] The nation needed more Polar heroes.

[67] Brekke & Egeland 1979:79
[68] Drivenes & Jølle 2004:118

Around the previous turn of the century Kåfjord was again an active local community, because the Copper Works had reopened in 1897, when the price of ore had risen. It became possible to cooperate with the management of the works, and the workers, about everything that had to be done to build an observatory on Haldde. The horses from the Copper Works were used for transporting building material and equipment.

The establishing of the observatory was exhausting work. During their first attempt to find a suitable mountain for their purpose, Birkeland and those with him were surprised by bad weather. They had to spend the night under a reindeer sledge, in -25°C, on the open plateau, and got frost bitten. Fingers were amputated, which resulted in an abrupt end to a career as a doctor for one of the assistants. The expedition was stopped, but Birkeland did not give up. The following summer he was back again and chose two mountains, Haldde and Talvik for further investigations. Two observation points, with a distance of three to four kilometres between them, would make it possible to take bearings or photograph the Northern Lights simultaneously from both stations, so that they could try and measure the height of the Northern Lights.[69]

Sem Sæland was to operate the station on the top of Talvik. Thus the world's first Northern Lights' observatory actually became a fact, and for the first time, four men spent the winter of 1899-1900 on the top of a mountain in Finnmark. It is said that the wind was so strong that it took three men to close and lock the door to the observation station. Unfortunately, that was not the end of accidents. In a snow avalanche 16th March 1900, near the Haldde observatory, two men lost their lives. The one was a local sea captain, Richard Lange, who often visited the observatory. The other was the 22-year-old student Elisar Boye, who was the youngest member of Birkeland's expedition.

[69] Devik 1971:32

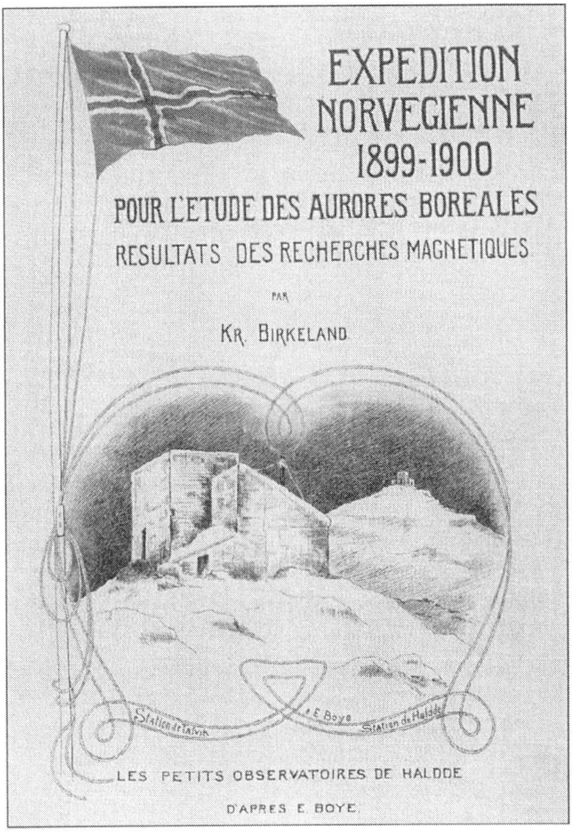


The front cover of Birkeland's dissertation with Elisar Boye's drawing of Haldde and Talvik observatory stations on the front cover. Archive: World Heritage Rock Art Centre - Alta Museum

The tragic experience left a lasting impression on Birkeland. He dedicated his completed doctorate dissertation to Elisar Boye. It was printed in French in 1901 entitled, "*Expédition Norvégienne 1899-1900 Pour L'études Des Aurores Boreales*". In a gesture of gratitude the front cover of the book was illustrated with a drawing that the deceased student had drawn of the observatory stations on Haldde and Talvik.

After processing the notes of this second expedition, it became clear that more observation posts were needed in the Northern Lights' zone, with, at least, a distance of 100 kilometres between the observation points. On the third expedition in 1902-1903 Birkeland chose Kåfjord as one of the four stations, where they would do simultaneous recordings of magnetic observations. The other stations were established on Iceland, Svalbard and Novaja Semlja. This time the observatory on Haldde was not used. Instead measuring took place in a small building down by the fjord in the village. Birkeland visited the station by the fjord twice during the winter, but his assistants were mainly responsible for running the station.[70] There must have been some trips up to the top of Haldde, as a photograph from that time show that, the then unknown, Roald Amundsen was on a visit to the observatory. The visit was because Amundsen was planning an expedition through the Northwest Passage, and he wanted the researchers to make instruments that he could take with him.[71]

Roald Amundsen on Haldde with Birkeland's assistants, Krekling and Sæland. Photo: Olmar Egenæs. Archive: World Heritage Rock Art Centre - Alta Museum

After the initial establishing phase the activities continued and research progressed forward, mainly at the university in Kristiania, but to a lesser degree in the Kåfjord region. Then in 1910, Birkeland took the initiative to make the Haldde observatory into a permanent station for Northern Lights' research and weather observations. He now believed that the top of a

[70] Egeland & Burke 2005:61

[71] Nielsen & Petterson 1993:28

mountain was better than the lowland to obtain, "the immediate changes in the air", and argued strongly that, "Haldde observatory would, in truth, be a shining scientific beacon for Europe".[72] In 1912 a new building was completed on Haldde.

Due to all that took place here, Kåfjord became well known for its high-quality research under the leadership of one of the country's foremost capacities in physics. In the village the activity on Haldde was regarded with goodwill. People were proud about what was taking place in their neighbourhood, and talked about the newcomers on the top of the mountain with respect. Those that had horses were more than willing to use them to transport wares and other consignments to the observatory, even though it was a demanding journey up the steep mountainside. Now and again the researchers made time to visit the Copper Work's people at the director's house. In a letter fifty years later the local shopkeeper, Laurits Heitmann from Kåfjord, writes:

> I think it was in 1900 that Professor Kr. Birkeland started to build an observatory, and I had considerable trouble in making sure that the necessary material was taken up to the impractical, and hard of hearing Professor, who was, nevertheless, very friendly. We had a lot of fun when he and Sæland, who was the leader of the Talvik station, came down to the Director's dance evenings, dressed in Sami footwear and "bonjour", (double-breasted overcoat), plus a belt and a large sheath knife. The humorous and friendly Sæland was often with us. Now and again we took a trip on the local ferry, as an amateur musical trio: Sæland tambourine, me violin, and my wife guitar. Sæland danced on the deck with his tambourine beating to our music, and the passengers had free Sunday entertainment.[73]

These events, which both cover serious research work, tragic accidents and amusing cultural meetings between the newcomers and the local people, created the foundation for the arrival of Dagny and Ole on Haldde Mountain in 1912.

---

[72] Letter from Birkeland to Sæland 23-09-1910

[73] Devik 1971:32

In Kåfjord dance evenings were arranged for those that worked at the Copper Works and the local people. Photo: Olmar Egenæs. Archive: World Heritage Rock Art Centre - Alta Museum

# CHAPTER THREE

# ALLIANCE WITH LOCAL PEOPLE

Waiting on the quay at Kåfjord. Photo: Olmar Egenæs, Archive: The World Heritage Rock-Art Centre Alta Museum

*A lot of Ole's photographs were thoroughly wet through. They were at the bottom of the wicker trunk.*[74]

In the early morning hours of 30th September, the couple arrive in Kåfjord. "There we were well received by the shipping agent and shopkeeper L. Heitmann, and spent the night there."[75] The shopkeeper had previously invited Ole to stay the night with him, and now they got a good night sleep before they started the journey up to the observatory.

[74] Letter from Dagny to her mother-in-law, 19.10.1912
[75] Dagny's diary from 1912

The shopkeeper Laurits Heitmann came to Kåfjord in 1898 at the age of 26, to be a trader for the Copper Works.[76] When the Copper Works were closed he bought quite a few of the buildings, and began to work for himself. That he was interested in music and was the organist at Kåfjord church[77] made him a central figure in the community, not just as a business trader, but also as a cultural person. The enterprising Heitmann built up quite an extensive family business, which consisted of the general stores, sub-postmaster, telephone station and steamship agent. He engaged boat builders from the neighbouring local authorities in Kvænangen to build the boat M/K Glimt.[78] With that boat he was able to transport both residents and guests, and not less, the doctor, midwife and priest when they had to be fetched from and returned to Bossekop.

Julie and Laurits Heitmann were among the acquaintances of the Krogness couple. This is a portrait of them when they were young. Archive: Private, Frithjof Heitmann

Heitmann was an important person for Ole and Dagny, someone who they had a close relationship with over the following years, both privately

---

[76] Heitmann married Julie Karoline Davidson and they had seven children.
[77] Heitmann (unknown year)
[78] Heitmann 2004:115

and as a tradesman. He sold them merchandise, made sure that all the products were sent up to the top of the mountain, and arranged for transport when needed. During the first period they were on Haldde they could not have managed without his help.

The warm welcome they received on their arrival at Kåfjord was in strong contrast to the harrowing time they had experienced on the boat. On meeting with the Heitmann couple they find good future friends, with whom they will have many enjoyable musical moments over the coming years. After the Krogness couple were installed in the house, their entire luggage was carried onto land, and some of their belongings were put into the Heitmann's house:

> Just think that when we arrived in Kåfjord the large black and grey suitcases and the wicker trunk were so wet at the bottom that we had to unpack everything and dry it. Mrs Heitmann was so kind and provided some underwear, almost all of Ole's wool and white shirts were wet through and had coloured 6 pairs of my knickers.[79]

It was not only clothes that had been packed into the suitcases. At the bottom of the wicker trunk were the photo albums with photos from Ole's and Birkeland's journey to Egypt, and now these were stuck together. The two of them had travelled to Egypt to study zodiacal lights, so that they could compare these, with what had been discovered at the Northern Lights' observatory on Haldde.[80] The album contained a number of photos from their experiences in that exciting country. The couple are aware that there will be a number of surprises when they unpack the rest of the equipment, when they get to the top of the mountain. There is not enough time for them to go through all the boxes during the short period they are with Heitmann. Most of the instruments, negatives, clothes, food, equipment and furniture will have to wait.

Whether the experience of sailing on a ship with a drunken crew was normal or not at that time, is difficult to verify. Like other parts of the country, drunkenness was a problem in Finnmark, and temperance work had been carried out for many years to try and get the better of it. Just a year previous to this there had been great uproar in Hammerfest, where the teetotallers had held demonstrations against the sale of wine after six o'clock in the evening. The town council had decided that the sale of wine

[79] Letter from Dagny to her mother-in-law 02.10.1912

[80] The Zodiacal lights are weak gleams of light in the sky, which are easy to see near the equator, a result of the sunlight, which is reflected from small dust particles near the earth's orbit. (See Egeland 1994:21)

should stop at that time, but the wine merchants meant that prohibition was against the law and angrily protested. The groups opposed each other and there was a very hostile atmosphere. When the Head of Police tried to calm the crowd, he was pulled down from the platform. Things went so far that the marine soldiers were brought in against the teetotallers.[81]

In Dagny's diary there are a number of references to the choice of wares in Heitmann's shop in Kåfjord. Archive: Private, Frithjof Heitmann

What we do know is there were other aspects that Dagny and Ole experienced on "Nor", which probably had happened to others before them. The boat that covered the route to Kåfjord and Alta from Hammerfest was often overloaded, which is documented in both the letters from Krogness and in historical publications.[82] The boat was originally used to transport post and passengers, and had very little room for all the goods they also had to carry. That "Nor" had been put into that route that day to replace "Alten", meant that it provided even less room for goods than was expected.

---

[81] Hølaas 1945:84
[82] Hansen 1999:5ff

It was the larger routes from Kristiania, Trondheim and Bergen that were economically lucrative for the Steamship Company, not the small routes that travelled into Alta fjord. The local routes were kept mainly to keep irritating competition at a distance so that nobody could take the routes from them. For the local population, it was frustrating that they were unable to get all the goods they needed from both Southern Norway and Europe.[83]

This episode brought with it a difference of opinion between Dagny and Ole, and Dagny writes this to her mother-in-law:

> Ole will not complain about them, because they will lose their jobs, and he does not want that. I can understand him, but on the other hand, if the First-mate had been steering the boat, and thereby sunk it with us on board, it is perhaps not surprising that one thinks that they deserved to be sacked, because in fact when we sailed at 2 a.m. it was pitch black. Ole says he has not got the time to meet in court. […][84]

Dagny wants to report the episode to the police, because in her opinion there was a danger of things going wrong. On the other hand Ole is not interested in reporting them, and his argument is that it was the sober Captain that steered the boat. Besides he does not have the time to complain now that he has just started his new job at the observatory.

The incident did not only annoy Dagny, but also the community in Alta and Kåfjord. Ole was pressed into writing an official complaint to the Director, Bull-Simonsen, in the Steamship Company (NFDS) in Trondheim. He did eventually write, but not until 12th November. Ole's excuse for the letter being rather late, is that, "[…] I first wanted to see whether these copies had been destroyed or not. Luckily it seems that they are useable."[85] His fright that hundreds of negatives with magnetic measurements could have been destroyed had been an absolute possibility.

---

[83] Hansen 1999:ff, Nielsen 1995:380

[84] Letter from Dagny to her mother-in-law, 19.10.1912

[85] Public Record Office in Tromsø: letter from Ole Andreas Krogness to Director Bull-Simonsen in Trondheim, 12.11.1912.

Ole travelled with Birkeland to Egypt in 1912. Archive: Private, Ole Andreas Krogness.

Ole adds further information to the letter concerning other things that were destroyed, among these the photo album that Dagny had made for him from his visit to Egypt with Birkeland. Dagny mentions this to her mother-in-law:

> The problem is that there are things that have been destroyed that are irreplaceable, and difficult to put a price on, for example the book with all the photos from Egypt became unstuck and the photos stuck together so they have gone to pieces, even though I put paper between each and every one of them.[86]

In his complaint, Ole goes on to say that the boat, "Nor", is too small to carry all the goods that are required daily on that route. He concludes the letter by calculating their loss to be between 50 to 100 Norwegian

[86] Letter from Dagny to mother-in-law, 19.10.1912

Crowns and adds, "that I definitely have not calculated too much".[87] The value was considerably much more.

Dagny later writes in a letter that she hopes the Steamship Company would be given a proper reprimand. It is not quite certain whether anything happened with regard to Ole's initiative. A later letter tells us that "Nor" is not running that route in the summer, when they would be travelling by boat again. It is possible that it was kept as a reserve boat, at least according to the shipping company's own bookkeeping, which had registered "Nor" as a reserve ship right up to 1915. "Alten" continued on that route up until the twenties, and on other routes in Finnmark right up to World War II. Most probably everything remained as it had been despite the letter of complaint.[88]

This is the prototype to the Krogness-camera, where one experimented with six glass plates and moved the lens between them. It was not a success, but was the origin of what is now known as the Krogness-camera. Photo: Ole Andreas Krogness, Archive: The National Library Photo collection.

[87] Public Record Office in Tromsø, letter from Ole Andreas Krogness to Director Bull-Simonsen in Trondheim, 12.11.1912

[88] Hansen 1999:54

Throughout the six years he lived on Haldde, Ole was often engaged in things that happened in Alta. There are letters where he uses his work situation to take up the region's problems and joys. When the new district doctor was sent to do war duty, Ole was indignant. He took his pen and wrote a letter on behalf of the people in Alta and Talvik rural districts, and argued that people in that area needed that doctor more than young healthy soldiers on a boat. He does not write about their need for a doctor on the top of Haldde, but no doubt it was a motivation for his engagement. Whether this letter gave any result, we do not know.

*To the Chief Administrative County Officer in Finnmark*[89]
*One doctor has covered the two rural areas, Alta and Talvik until a few months ago when it was changed to one doctor for each region. However a couple of days after the newly announced doctor in Alta, Gjessing, arrived there, he was called up to do duty for the Norwegian neutrality service on board the Coastguard ship, which was stationed in Vardø.*
*A couple of days ago he returned, but we believe he will be called back again quite soon as the marine medical service has hardly any doctors.*
*After consulting with the doctor, we have found that his service on board consists of him giving medical help to a crew of about 150 men. If necessary a medical examination is to be done in advance so that all the men are healthy.*
*Moreover the coastguard boat is never so far away from land that a doctor could not be contacted within 6 hours. There is also a medical orderly on board, who could give the necessary first-aid in a possible emergency.*
*Therefore we find it quite strange that a doctor, without further notice, can be removed from a district such as Alta, which is the country's largest medical district, and where his proximity is absolutely a necessity.*
*Whether the circumstances in general are such that there is a need to call a doctor away from his district to be used for such military service without the provision of a temporary doctor to take over, is something which we will not go into, but we would like to point out that in this special situation, we can only describe it as being irresponsible with regard to the long mountain journeys that the doctor must take from time to time.*
*Such a journey takes at least 8 days, and if there is only one doctor on call for both the Alta and Talvik regions, it means that one is simply cut off from medical help.*

[89] It is not known which year the letter was sent, as the number of the year can no longer be read. It was probably 1914 just after the First World War started. The date on the letter that we can read is 19.01

*The nearest doctor is in Hammerfest, and will only rarely have a chance of travelling to this area, not to mention how much such a visit would cost. Therefore, we believe that it is essential that there is a doctor in each of the district areas of Alta and Talvik at all times, so that in emergencies the Talvik doctor could deputize for the Alta doctor when the journey, for example is, to Kautokeino.*

*We take the liberty of requesting the Chief Administrative County Officer, to try and prevent the district doctor in Alta from being called away from his district in the future, to do duty in the Norwegian neutrality service, as has been previously mentioned.*

*Yours respectfully,*
*In accordance, with the staff at Haldde Observatory,*
*O Krogness*

The view over Kåfjord with the church on the other side of the bay. Archive: The World Heritage Centre for Rock-Art – Alta Museum.

# CHAPTER FOUR

# EARLY DAYS ON HALDDE

Goods and luggage were transported like this on a sledge. Sometimes selected passengers also sat on it, especially when they came to the steepest incline. Archive: World Heritage Rock Art Centre - Alta Museum

*I really like being here very much, and have so much to think about that I am never idle. Today I have moved the sideboard into the house. It had to be stained first and two pieces that had fallen off had to be glued back into place.*[90]

Dagny and Ole are now almost at their journey's end, but at the beginning of their life together. It is a cold but sunny day as they continue to climb up the mountain. It is covered in snow and this has been the

[90] Letter from Dagny to her mother-in-law, 31.10.1912

deciding factor for how they should travel. Their luggage was placed on a large wide sledge, which was pulled up by a horse. According to the calendar it is the 1st October 1912.

It soon became steep and continues upwards for the first hour. Soon they could see the beautiful view. After climbing for another half an hour the view was even more magnificent and impressive, and they could now see right out to the sea, Kåfjord and on towards Alta. Sæland had told them about his experience when he did the same journey with Roald Amundsen, some years previously, but then on the way down from the top of the mountain:

> But in the dusk when we went down again, it seemed as if the atmosphere from the mountain and the plain below took a stronger and stronger hold over us. Where we walked it interchanged between snowdrifts and mountain gravel, and at times small carpets of yellow violets. Then we arrived at a high ridge from which we saw the whole of Alta fjord and the Alta valley opened up in front of us. A long way to the south east, within Porsanger fjord, the Gaisserne mountains shot up over the plains like a dream palace in the evening sun, and southwards the plain bulged like a never ending congealed sea. We had walked in silence for a while. Then Amundsen stopped and took a long look all around. He then came with some words spoken quietly, and which I have remembered ever since and will always remember in connection with him, "What a country," he said, "it is so lovely that one is almost lost for words."[91]

In 1912 the group move on after gazing at the view upwards and over the ridge of a gentle slope. Here the terrain flattens itself out, and the landscape stretches like a white plateau towards the base of Haldde Mountain. Dagny can now glimpse her new home for the first time, as two small blocks in the distance. There is little vegetation and naked mountaintops, gold, exotic and beautiful at the same time, and very different from the luxurious vegetated mountains in the eastern part of Norway.

The easiest part of their journey is now in front of Dagny and Ole, and on their way over the plateau, they pass the dark entrance to one of Kåfjords closed down copper mines. Dagny described it like this in her diary, "[…] the road began quite steeply, but became more even higher up, until one reached the last bit up to the top of Haldde. It is extremely steep, so that I was put on a sledge, and pulled up by a horse that was harnessed to it."[92] As she rounded the last corner on the steep incline she must have

[91] Devik 1971:65

[92] Dagny's diary 1912

seen the house, a house built of stone and partly covered in snow, towering in front of them. She must also have caught a glimpse of the observatory behind the main house, soaring high up almost as if it was a continuation of the mountain, an enormous beacon towering over them.

View seen after the first ascent up to Haldde. Archive: World Heritage Rock Art Centre - Alta Museum

## The Buildings

There are sheer drops on three sides of the observatory, and on the fourth side a steep slope dives down towards the main-house. The observatory is a solid 36 square metre building with one metre thick walls.[93] The front door is placed so that it is sheltered from winds and weather. The assistants are to live and work in the room on the first floor,[94] and will be kept busy running up the stairs leading to the actual observatory tower, which has a bricked-in parapet. Up here it is possible to see over the mountain peaks and catch a glimpse of the fjord in the distance, a spectacular view and an excellent place for nightly observations. The parapet is probably the reason why the building became known as the Castle.[95]

Ole had to carry ice and snow inside, from the beginning of their stay, to be melted for the water supply. Archive: National Library, Photo collection.

[93] Mentioned in both Krogness' letter to NVE about Kåfjord's hydro-electric power station, 24.09.1925 and Alta Museum's notes concerning the uses and restoration plans for the Northern Lights' observatory on Haldde 1999.
[94] Earlier used for people staying overnight, for example Roald Amundsen stayed there when he visited Birkeland on Haldde.
[95] Both the observatory and the house in Kåfjord were called "the Castle", in a number of journals, which could lead to some misunderstanding. To evade such a misunderstanding we always use the place-name as well as the nickname when it is not given in the text.

Between the main house and the observatory there is a tunnel, which the researchers can use when observations are to be carried out, and it is bad weather. Apparently it is both practical and exciting, but Dagny experiences it quite differently. "I was with Ole yesterday and examined the underground tunnel, but it was horrible, steep and uneven and unpleasant to walk through."[96]

The main building is built of solid local stone and is one and a half floors high. The house is placed in a north-south direction and the entrance is on the long side of the building facing east. On the first floor there is a kitchen placed on the southern side, bedroom on the north side and the two living rooms next to each other. The attic is airy and big with a guest-room, and in addition there is plenty of room in the cellar. Dagny writes excitedly about the house in a number of letters to her mother-in-law, and tells her about the interior and the placing of the furniture they have taken with them, which her mother-in-law probably knows. "The white cupboard is charming, but could easily have been twice as high, as the builder has raised the ceiling higher than is shown in the plan drawn of the house, so it is more roomy than we thought."[97] Like others, who move into their first home, Dagny is enthusiastic and busily occupied with installing things and making it cosy to live in. "I try very hard to help, but unfortunately my back is unable to tolerate very much. It is so lovely being in our own house, and the bedroom is tremendously comfortable. [...] You must come up and visit us because the house has been organised magnificently. There is a large cellar and huge attic."[98] Her worries that the house would be too small soon vanish.

Dagny has taken what is necessary with her for establishing a conventional middle-class home. Plush chairs, sideboard, dining table, silver ware and crystal glasses; they all find their place in their new home a long distance from everywhere and everyone. Not least, a piano will also soon find its place there.

---

[96] Letter from Dagny to her mother-in-law, 19.10.1912
[97] Letter from Dagny to her mother-in-law, 19.10.1912
[98] Letter from Dagny to her mother-in-law, 02.10.1912

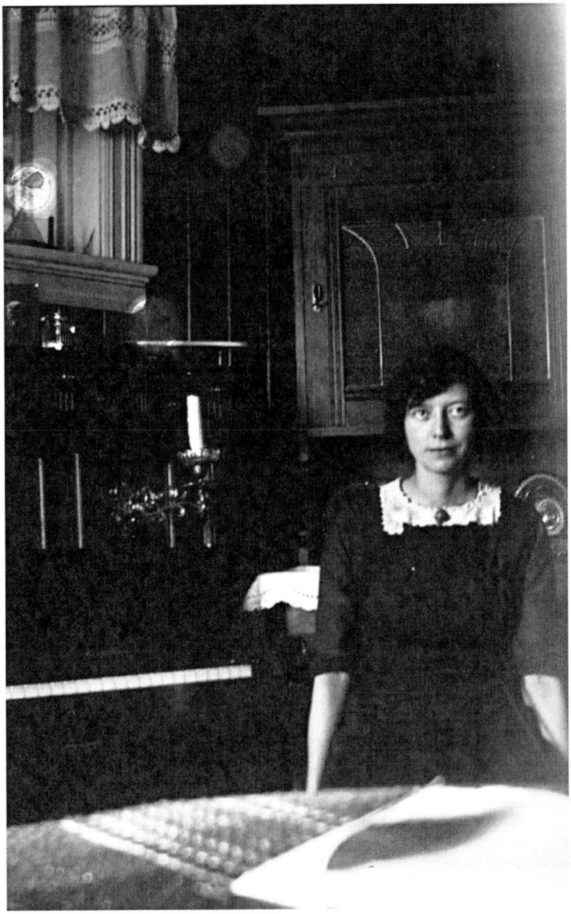

The young Dagny in her new home. Photo: Ole Andreas Krogness. Archive: National Library, Photo collection.

In the beginning Dagny uses most of her time to get to know the house. As she gets more into a daily routine, she discovers that not everything about the building is as good or practical as she had first thought. The house is large and fine, and has all the room one could wish for, but Dagny gradually sees that the building priorities that had been made have not been thought through well enough. She writes to her mother-in-law on 19th October that even the people in Alta reacted to it. "Our bedroom is unfortunately facing north and the kitchen south, but we leave the door to the dining-room open so that we are able to let a little bit of sunlight in."[99] It is the opposite of what one would wish for in a house, as the kitchen should be cold towards the evening and the bedroom full of sunlight during the day. She is frustrated that the bedroom has been placed on the north side of the house, because when she wants to air the room she has to do it through both the living room and the kitchen. What happens is, that all the heat disappears from the whole house.

The contemporary handbook for expectant and newly fledged mothers that Dagny reads is entitled, "*Mor og Barn*", a book for mothers and children, written by Doctor Greve,[100] and there it, "[…] recommends that future mothers should have a bedroom facing south to let in as much light as possible."[101] Other advice that is given in the doctor's book is about fresh air. Dagny reads:

> The pregnant mother breathes for two and must have fresh air. During the summer you should therefore be outside, and in the winter both before and after dinner should take a regular walk of 2 hours, despite what the weather is like. All kinds of work outside is healthy.[102]

It is impossible for Dagny to take long walks, as the weather is still too bad and the ground too difficult to walk on. There is very little room outside the house. Dagny finds out what to do on the days when the weather is at its worst. She puts on her "Icelandic jumper and Sami boots, and with my head well-covered, I stand in the doorway to the hall entrance and breathe as deeply as I can, because outside the storm is blowing so furiously that it is not a good idea to put one's head out".[103]

Even though the placing of the bedroom irritates Dagny, there are other challenges she discovers that to a greater degree influence her daily work. The kitchen, the heart of the house, is the most important room for her and her maid to work in. As the wife of a researcher, who has two assistants, First assistant Aabakken and Second assistant Hoel, she is responsible for providing food for them. She discovers that the kitchen stove does not have enough room for all the saucepans when she makes meals, and she appoints the workmen to make extra metal plates to stand beside the actual stove:

---

[99] Letter from Dagny to her mother-in-law, 19.10.1912

[100] Greve (1906) the doctor author referred to in Dagny's letter to her mother-in-law 03.05.1913

[101] Greve 1906:1

[102] Greve 1906:1

[103] Dagny's diary from 1912

> You ask what the workmen have been doing, everything under heaven and earth, I have two cupboards upstairs made out of the instrument boxes, a baking trough and different metal plates beside the kitchen stove, because it is so small that the saucepans fall onto the floor when I have some stewing pans on it, and so now I have room to put the primus.[104]

Dagny also questioned other parts of Birkeland's planning. She finds it strange that anyone could think that a maid should sleep on a plank bed in the kitchen. Maybe it was usual in previous times, but for Dagny it is unthinkable. She feels that the kitchen is small enough as it is, and absolutely not a place for a young girl to live. Living together and being cooped up in a house for most of the year, demands that living conditions are good for all those living there, including the maid. It ends with Dagny giving the maid the large guest room in the attic. This means that they no longer have a room for visitors, or for those who have to stay overnight for other reasons. The large house has now suddenly become smaller.

## Piano, Music and Party

While Dagny has made sure that the crystal and the furniture has been safely transported up to their house on Haldde, Ole has concerned himself with something that makes him happy, musical instruments. Starting a research station is demanding, and a bit of music can now and again be absolutely necessary. That all went well in the transportation of the violin, viola and cello up the steep ascent, is perhaps not so surprising, but how he managed to get a heavy and unwieldy piano up there, is a question many will wonder about.

The piano was sent up from Kåfjord a month after the family, and arrived on 31st October. Ole had visited Haldde before they moved there, and knew how difficult it would be for those who would have to carry such an instrument, especially the journey up from Kåfjord. He had very carefully planned the transport beforehand, and his mother was involved and had early reports. In her letter to her mother-in-law Dagny commented that, "the piano will come up in winter conditions".[105] The couple had thought it would be very expensive to transport the piano. They were given an estimate of 50 kroner[106], which would be a good 2.800 kroner today. They got away with paying 17.50 at today's value of 900 kroner.[107]

---

[104] Letter from Dagny to her mother-in-law, 29.11.1912
[105] Letter from Dagny to her mother-in-law, 02.10.1912
[106] Letter from Dagny to her mother-in-law, 10.11.1912
[107] National Bureau of Statistics 2013

The newly married couple had music as a common interest, and on their living room wall over the piano they had the death mask of Beethoven. Photo: Ole Andreas Krogness. Archive: National Library of Norway, Photo collection.

The house is placed in wild and beautiful terrain with a wide view all around, ideal for a Northern Lights' researcher. Archive: National Archives. of Norway.

To be able to transport the piano up to the house Ole realised that he had to do something drastic. It would be impossible to put a piano on a sledge. He found that the best way was to take the piano to pieces, pack the parts carefully in boxes and transport a portion at a time.[108] Once all the parts had reached the house Ole put the piano back together again, and then tuned it ready for use. Not an easy job for an amateur, but clearly something he mastered. For Dagny, Ole's aptitude for music is a source of great joy. "Yes we will have lots of good times with the piano keeping us company",[109] Dagny writes to her mother-in-law.

Music together with food and drink was from the first moment an important social part of their daily life. Dagny has learned to play the piano a little bit, and does her best when she accompanies her husband when he plays his cello. There were no neighbours to hear them, and it is unknown if any sound sieved out through the thick stonewalls. The picture of the newly-wed couple sitting in their living room high up on an isolated mountain, playing for each other, while a storm bellows round the walls of the house, is to put it mildly, rather appealing.

After a few weeks they have finally managed to get things into place and decide that it is time to have a party for the Haldde colony.

> Yesterday we had a big party, for dinner we had soup á la reindeer and boiled cod with red wine. Ole held a short speech for the occasion, and with the coffee we had cake from Aunt Anna Moe, which was sent with the grocery to the shop here from Strøm (a shopkeeper in Trondheim).[110] I have so much cake and will try and put some aside for Christmas, although Ole, as you can imagine, wants to eat all of it now.[111]

After dinner they continued their social activities such as playing cards and of course music:

> Yesterday we initiated the living room with a party of bridge players and "jogasta" that is distilled spirit in Sami. Here it has become extremely cosy now that all the cushions are on the sofa, pictures on the walls and curtains for the windows. Ole and I played the cello and piano together, but Ole is a bit impatient about my rhythmic beat. Yes, I am apparently hopeless, but things will get better. Aabakken has a good singing voice so that he and Ole have begun to sing Gluntarna together.

---

[108] Reported in a newspaper cutting owned by the family. Unknown newspaper and date 1934, by Mathilde Mathisen

[109] Letter from Dagny to her mother-in-law, 20.11.1912

[110] Shopkeeper in Trondheim

[111] Letter from Dagny to her mother-in-law, 29.11.1912

They often had visitors up there. Photo: Hilding Köhler Archive: World Heritage Rock Art Centre - Alta Museum

Dagny liked having visitors, and a social setting, also when they can keep the party going until the early morning hours. By chance the Rev. Ording from Hammerfest visited Kåfjord church, and after the service Heitmann took him up to Haldde, and he stayed with them overnight.

> The Rev.Ording plays the violin and as you can imagine there was music here. Ole and the vicar spent the whole night playing, we went to bed at 6.30 a.m., apparently that is quite usual here in Finnmark. The vicar played the violin quite beautifully, and he sang a bit and also played a bit on the piano.[112]

## Stormy Height

Ole knew about the bad weather on Haldde from his previous visit, and now Dagny experiences what she has been told. Not only did her need of fresh air lead to other routines than daily walks outside, but the weather also influenced almost all their activities in daily life.

[112] Letter from Dagny to her mother-in-law, 08.01.1913

The weather lightens up after a day of snow and wind. From the flat roof on the washhouse, snow is collected to melt for the water they need. Archive: World Heritage Rock Art Centre - Alta Museum

"Here the storm has continued for the whole week and they could not use the funicular. Yesterday the wind speed was 30 metre so that in the evening nobody could stand upright."[113] The weather often prevented them from going to church on a Sunday, but Dagny writes that it is so cosy inside the house that she is not bothered. Dagny continues to write about the weather throughout the winter months, and at times it is very violent.

> Yesterday night the wind blew so hard that the smoke hat on the chimney pot had blown off. Ole went out and found the bits in different places, one was also blown off "The Castle", but that one had gone over the edge of the steep incline, and it was impossible to find.[114]

The weather changed and the sun and moon appeared as soon as the clouds disappeared. This was the first winter with a period of darkness that Dagny experienced, and this polar night affected her:

> We have had some lovely weather here for a while both moonshine and a glimpse of the sun at times, but now the sun has disappeared and will be away for a long time. The Polar night has arrived, and it is not light until

[113] Letter from Dagny to her mother-in-law, 19.10.1912
[114] Letter from Dagny to her mother-in-law, 10.11.1912

> about 10.30 a.m. and at one o'clock I have to light the lamps, at least in the kitchen. Ole is now getting up at 7.15 a.m. as well as the others, unless they have overslept, as there is so much measuring to do, three times a day, of the wind, weather, snow and clouds.[115]

Much is new during these first months, among other things Dagny experiences that the journey down to Kåfjord can be tough as the weather can suddenly change. Therefore she does not want to go with Ole when he wants her to go to the market in Bossekop. Even her mother-in-law advises her not to go.

As the time for the market gets nearer, Ole has to work hard to persuade Dagny to accompany him, and in the end Dagny gives in. She experiences the market as being completely different from anything else she has been used to from southern Norway. It is a traditional gathering place for people to meet when the Sami people come to sell their wares.[116]

> In the evening, at 8 p.m. I decided to go, and had to begin to put out the skirt I use to walk in, and prepare some food for Hoel, who would be at home on his own as the maid had already gone [...] In Bossekop it was terribly cold, minus 20 degrees, we wore fur coats and I had my Sami footwear, "skaller", so that only my face froze, but I was glad to get home again. The last bit from the storage-house I had to walk and the slope is vertical with snow crusted on it so that Ole had to lead me and stop every other step.[117]

They bought a lot and Dagny writes in her accounts that she has bought wares for 50 kroner from the market.[118]

Despite wind, snow and ice the young Dagny has many lovely memories and impressions. One of them she writes poetically about in her diary:

> So we have days with strong blue air and blue-white colours over the mountains and fjords way out in the distance, while the sun tries to give us a glimpse of its pure golden rays, but is too weak now. We also have clear starry night skies or moonlight ones. First of all we had red-gold climbing up to the blue heavens, but now it has turned pale with its death mask looking outwards over the silence. Perhaps it is most beautiful when the Northern Lights have reserved themselves for one evening, as then lights shine and flicker in yellow, red and green over the heavens.[119]

---

[115] Letter from Dagny to mother-in-law, 29.11.1912

[116] Nielsen 1995:280

[117] Letter from Dagny to her mother-in-law, 15.12.1912

[118] Dagny's diary from 1912

[119] Dagny's diary from 1912

Her sælges Ferdigklæder &
Underløi & Trekol
Til Billigste Priser.

Bossekop market was an important meeting place for people in the region. Photo: Olmar Egenæs. Archive: The World Heritage Rock-Art Centre – Alta Museum, [Photo top left], and Norwegian Museum of Science & Technology

# CHAPTER FIVE

# THE ENIGMATIC GENIUS

Birkeland outside the University in Kristiania. Photo: Carl Størmer Archive: Norwegian Museum of Science & Technology

*Yes I must say that it is marvellous that he remembered us.*[120]

Kristiania $19^{th}$ Aug. 1912

Dear Krogness,
I have received your letter and send it back with endorsement. Give me a total specification, (total sum Kroner.), of what has been ordered to date. I have ordered a good spectroscope myself. Say to Nippoldt that I am charmed that he will come and see the experiments; they are now absolutely magnificent. But I must be there myself, so I will write later so

[120] Letter from Dagny to her mother-in-law 29.11.1912

that we can agree upon a time. I have recommended Aabakken (?)[121] and Hoel as 1st and 2nd assistants, so they will be appointed as soon as possible. I am sending the money at the same time.

With Kind Regards to you, and your wife. Kr. Birkeland.

This short letter, sent to Ole while he was in Germany, tells us something about what kind of person Birkeland was, concise and directly to the point! This was clearly shown as he sent the long nine-page letter that Krogness sent to him, in return with the "necessary endorsement". The endorsement was on the first page of the letter, with for example a "yes!" in the margin as the answer to Ole's well-grounded arguments and questions, and the numbers 6 and 4 on the actual text where Ole suggested the different number of instruments he needed to buy.

Another glimpse of Birkeland is related by Sem Sæland, who was his assistant for many years, in a radio lecture from 1936. He tells about seeing Birkeland on 17th May 1900; a little stocky, well-dressed, mild and cheerful man with brilliant blue eyes behind gold-rimmed glasses, and a hat down on his neck. He was in continual motion carrying a huge bag of confetti. Suddenly he sprinkled it over the most beautiful woman he meets, and then he quickly throws a hand-full of the fluttering, colourful small pieces of paper into the face of the sourest and the most ceremonious man he meets, especially his own colleagues. Birkeland was then 33 years old and had been a professor for three years.[122]

Ole's collaboration with Birkeland went back to when he was a student at the University in Kristiania. While he was his student Ole had accompanied Birkeland on a research trip to Egypt to study the Zodiacal Light.[123]

[121] Birkeland seems to be unsure whether he has written his name correctly.

[122] Sæland 1936:33

[123] About the Zodiacal Lights, see footnote 80.

Student and Professor on fieldwork in Egypt 1912. Archive: Private, Ingrid Løvlie

Birkeland discovered quite early on that Ole was a promising student and was obviously so pleased with him, that he involved him in his projects. He had made no mistake. In his final examination Ole got a top-grade in mathematics. Olaf Devik, his study friend, achieved the same result. Two such excellent students were confidently the best men to continue with research work, and Birkeland made sure that they were given good working conditions within his own field of interest.

During the period on Haldde contact with Birkeland was necessary, as he was responsible for financing the research activities, both by applying for public funds and from his own private funds. He also contributed to making life comfortable for his employees from his absent position. Birkeland showed great concern for the married couple that he was responsible for placing on the top of a mountain in Finnmark. This was made abundantly clear when they arrived at the house on Haldde and found that he had bought blinds for all of the windows.[124] The probable reason for this was that Birkeland had discovered for himself that it was necessary to block out the light from the midnight sun for a number of months each year if one was to be able to sleep. It was known that he struggled with sleeping problems for most of his life.

Dagny reported, in a letter of 29th November 1912, that they have received a telegram from Birkeland, which stated, "Have given 50 kroner

[124] Letter from Dagny to mother-in-law, 02.10.1912

(ca. £2,40[38]) for Christmas sweets for the observatory." She is enthused over him remembering them and tells her mother-in-law that she immediately ordered, "nuts and marzipan from Heitmann and later they would get a box of oranges, because she believed that just then the oranges were too sour and expensive."[125] She had to be careful with her use of money. Everything had to be reported. The price of food she bought was recorded in the accounts, which Ole kept and sent to Birkeland and the Government. The accounts included everything in connection with the building, housekeeping, wages for fellow workers, both research assistants, maids, transport, work people and the cost of purchases and any other necessary services required. From these accounts and letters to Birkeland we see that a great number of the posts in the accounts, concerned instruments that had to be ordered and transported from quite a distance.

H. Devold (1940) claimed that nowhere had it been registered, that the wind blew worse than it did on Haldde. The registration was done with the equipment shown in the above photo. Archive: National Library of Norway - Photo collection.

---

[125] Letter from Dagny to mother-in-law, 29.11.1912

The letters from Ole to Birkeland are very detailed. They have obviously agreed that the Professor himself should be responsible for the research activities, and the observatory's Director had to comply closely with his instructions, as implied in a quotation from one of Ole's letters:

> As you can see I have made a separate account for each worker, which has resulted in quite a lot of paper being used. The reason for this is so that the workers are unaware of how much each of them earns.[126]

Birkeland was known to be rather careless with his accounts and book keeping. When the Auditor General of Norway, in a letter asked about some vouchers, Birkeland replied quite simply in a letter with the remark, "I remember that. Kr. Birkeland".[127] Sometimes he did inadequate preparatory work and afterwards muddled up the receipts.[128] Yet he was thoughtful enough to order food for the house on Haldde. On the other hand things were not always of the best quality. Dagny became aware of this when she started her housekeeping. She was indignant and disappointed with the shopkeeper and wrote to her mother-in-law about this, "and the potatoes he had sold to Birkeland at 7.50 kroner (ca. £0,36[38]) per barrel were half rotten and soft, we have to sit for an hour just peeling potatoes".[129]

Even though Birkeland lived quite a distance from them, he involved himself in their reports about their daily life, and the research reports from Haldde. He had remarked that, "I think it is best that the observatory is under my jurisdiction as long as I live".[130] An example of the contact with Birkeland can be found in one of the many letters to, "*Kirke-og Undervisningsdepartment*," the Ministry of Church and Education, dated 8th October 1914, where Ole reported about their investigation, and results of their work. Among other things he wrote:

> In this connection I can also mention that Professor Birkeland, has just succeeded, with regard to this investigation in getting Greenwich Observatory to send us special telegrams showing sunspots.

The observatory on Haldde gave Norway international fame as world leaders in their research of the Northern Lights. Birkeland's investigations

---

[126] Letter from Krogness to Birkeland, 18.11.1913
[127] Devik 1971:37
[128] Drivenes & Jølle 2004:122
[129] Letter from Dagny to her mother-in-law, 10.11.1912
[130] Birkeland's letter to Sæland 28.09.1910

in the Nordic region had given results, and he had concluded that there was a significant connection between terrestrial magnetism and meteorological conditions. Therefore he believed that a permanent magnetic-meteorological observatory should be established on Haldde.[131] Unfortunately after the first years' intense work around 1900, things stood still on the mountain for some years, and Birkeland worked with other projects.

In 1905 he married Ida Augusta Charlotte Hammer. Birkeland's private life is not mentioned in either Dagny's or Ole's letters. Information about any private life is mainly restricted to work tasks that are carried out daily. Rarely is there any mention of the intimate or emotional side of their life, even in Dagny's letters to her mother-in-law. Jago's book, "*Nordlysets gåte. Beretningen om Kristian Birkeland, pioneren som ofret alt i vitenskapens navn*", is an account about Kristian Birkeland, the pioneer who sacrificed everything in the name of science. The author mentions Birkeland's marriage as not being so successful. Research and other professional activities took up most of his time and energy, which affected their married life rather negatively. Most of his attention was concerned with the many experiments he carried out, and discoveries that were a result of these. This gave him the opportunity to profit, increase his private capital, and thus finance further research.

In 1906 he engaged Lars Vegard and Ole Andreas Krogness, who at that time were students of science and mathematics, to work with the large amount of material concerning magnetic storms, collected from the Arctic expeditions. In that way Ole became an important contributor to Birkeland's great work, "*The Norwegian Aurora Polaris Expedition 1902-03.*"[132] Vegard, at a later date, became a well-known and prize-winning researcher.

Olaf Devik, who together with his brother Karl, also became assistants to Birkeland, described the job as being quite different from anything they would have experienced if they had worked for other professors:

> There was no routine work that Birkeland specifically asked his assistants to do; they were immediately included in discussions, worked quite freely and independently and experienced what it was like to be stimulated by working with others as well as being independent in a powerful field of study, which constitutes a scientific environment.[133]

---

[131] *Værvarslinga for Nord-Norge 25 år, Festskrift*, 1945. (A memorial volume).

[132] Pages: 319-438 about magnetic storms and p.757-801 about Currents (ref. in Devik & Krogness' application papers for the position of professor in Bergen).

[133] Devik 1935.

The Birkeland couple had a fashionable home in Kristiania. Photo: Severin Worm-Petersen, Archive: Norwegian Museum of Science & Technology

Birkeland's laboratory at the University in Kristiania was so fully packed with research equipment that only those who worked there on experiments, were allowed in. If any students had questions they had to stand in the doorway and ask him from there. The laboratory was chaotic and noisy with lights flashing and strange smells. The assistants that worked there had to be very careful. A number of them got electric shocks, but none of them were seriously injured. All of them had to work with one hand in their pocket, so that the electric current did not go through their heart.[134] Olaf Devik wrote later that during the years 1903-1913 Birkeland "burnt his candle" at both ends, so it is probable that the hard work affected his health.[135]

In 1910 the energetic professor was again on Haldde with his instruments, this time together with Ole Krogness to study the Halley Comet. It was supposed to be visible that year on 18th-19th May. The two of them stayed in the old primitive observatory that had been built ten years earlier. They defied predictions of the world's destruction that many feared would be the result if the comet collided with the earth. Astronomers had

[134] Egeland & Burke 2005:35

[135] Devik1976: 38-39

discovered that the comet emitted cyan, a chemical that was used to make prussic acid. People read warnings in the newspapers that astronomers believed that all living things on earth would be wiped out. The world could expect that many people would be poisoned, and there were many prayer meetings.[136] Researchers took the whole thing quite calmly.

The observations they made supported Birkeland's belief that a permanent observatory on Haldde was needed, and he began to work to make this possible. His intentions were met with opposition from Professor Mohn at the Meteorological Institute in Kristiania. Mohn preferred that they established a main position in Bossekop, a more densely populated area between Kåfjord and Alta, and had a small station on Haldde. His reason for this was that people would not want to live in such a remote place. In a letter to Sem Sæland, who had himself been on the first expedition to Haldde, Birkeland writes about this. His conclusion is:

> I believe that once one has a solid waterproof timber house, there will be no danger of that. It only takes four hours to get down to people, and the staff at the observatory can take it in turns to have eight days at a time on duty, if that becomes difficult there is always the possibility of using the telephone. As long as I live, I will guarantee that we shall not be without people up there. And if we poor Norwegians use so much money on research, then the consideration of physical or spiritual exertion must not obstruct us from choosing the best.[137]

As far as Birkeland was concerned research should always take first place. It was the main task, one that demanded much of those that participated. At the same time he was very considerate when arranging the accommodation for those who would be living on the top of the mountain, by providing a house down in the village which they could make use of. It was a neighbouring house to "The House". History tells us that Birkeland got his own way. An application from Birkeland to The Royal Ministry of Church and Education, was sent the same year.

> I think that a permanent observatory on Haldde will be a goldmine for scientific research, and will especially be of use to meteorologists, and weather predictions will be of great practical significance for fishermen in the north. An observatory there, at 70° latitude will be without any contradiction a scientific station that has all the conditions required for making it a special prominent, not to say, a most outstanding place for

[136] Hølaas 1945, newspaper report in one of the Norwegian liberal newspapers (*Tidens Tegn* or *Dagbladet*) 13.01.1910

[137] Egeland 1994:18

geophysical observers all over the world. Here is a possible non-scientific aspect, which may carry some weight in the decision of those in authority. An experimental permanent institution in Finnmark, established by Norwegian scientists would contribute to asserting the importance of the most northern part of Norway, internationally. In turn this would have an effect on the local population and their national identity.[138]

The House, also known as, "the Castle", was for the researchers' disposition in Kåfjord. Archive: The National Library of Norway- Photo Collection.

In his application we can see that Birkeland argues practically, and pragmatically that an observatory would be beneficial, with its weather warnings for North-Norwegian fishermen. He also believed that it was crucial for Norwegian authorities to have such an institution, as it would encourage citizens in the far north to feel valued members of a Norwegian society. The implication behind this is that the country needed to unite people, and strengthen its position as an independent nation, after the dissolution of the union with Sweden. Birkeland was an adherent supporter of Norway, and believed it should look after its own interests, also while it was united with Sweden. His political engagement persisted, and we see that it is part of his strategy for getting his cases accepted.

[138] Devik 1976:8 and, "*Varvarsling for Nord-Norge*", 25 år, Festskrift 1945:14. (A memorial volume, to celebrate 25 years of weather forecasting services.)

Applications were endorsed with particular emphasis on weather forecasting services.

Throughout his academic career Birkeland travelled a great deal, and made a great many international research connections. In 1890, when he was 23 years old, he completed his academic education in Science. When he was appointed to a Ph.D. position at the University in Kristiania, he received a travel grant, and lived in Paris, so that he could study Maxwell's theory on the magnetic field. After that he travelled to Bonn and to Genève to meet leading scientific researchers in his field of study. Later he spent long periods engaged in research in Egypt and Japan.

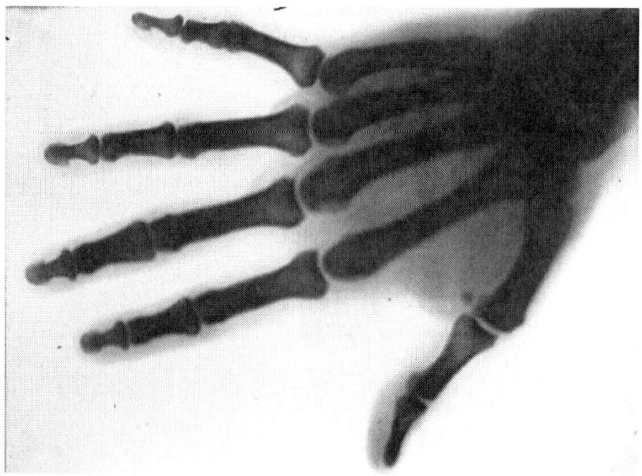

The first X-ray taken in Norway. Photo: Kristian Birkeland Archive: Norwegian Museum of Science & Technology.

Birkeland was an active and busy researcher, who today is regarded with great respect, by those engaged in the same field of study. He published about 70 large scientific works, the majority of them in French, and he wrote three monographs, one of them over 800 pages in length. On the 6th October 1898, when he was 30 years old, he was appointed Professor of Physics.[139]

He was also an industrial researcher and developed 59 patents. Twelve of them were connected to what was called the, "*the Birkeland-Eyde-Furnace*", and was in fact the start of "*Norsk Hydro*", one of Norway's absolute largest industrial concerns. What is interesting about this is that the Birkeland-Eyde-Furnace, goes back to the first winter spent on

[139] Egeland 1994:6

Haldde, when the group on the mountain were invited to Anders Quale, who at that time was the Director of the Copper Works in Kåfjord. Quale was Swedish, and had moved north to the Copper Works when the mining industry became, once again, financially worthwhile. What interested Birkeland, was that Quale had just finished building a hydroelectric plant with a power station under a waterfall in Kåfjord. That particular waterfall had a 370 metre drop, a good 70 metres higher than the Eiffel tower, something that impressed Birkeland. On that degree of latitude the plant was a noteworthy piece of engineering skill, and Birkeland wanted to see how it functioned in practice.[140] The water dropped down hundreds of metres through steel pipes to massive turbines that transformed the energy created by the water into electricity. Quale had problems with the mechanism for turning the current on and off, as quickly as possible. There was no controlling element that could do this. Birkeland, who was already interested in using his scientific knowledge to develop machines, decided that he could contribute to the development of the hydroelectric industry, and in that way place Norway on the map of industrialized nations. If he invented a power switch, that could do this job, he could earn some money, which he could use to finance more research on the Northern Lights.

Back in Kristiania, Birkeland began to experiment with the development of a power switch. It was during this experimental work that Birkeland discovered a beam of light, which later could be a decisive factor in the production of the air's nitrogen to saltpetre, which could be used as artificial fertilizer. Through that work he came in contact with the hydroelectric industry. In 1903, Birkeland and the industrial tycoon, Sam Eyde, met for the first time at a dinner party, held by the then secretary of the Ministry of Agriculture, Gunnar Knudsen, a member of the Liberal Party, later Norway's Prime Minister, in 1908-1910 and 1913-1920. Birkeland was also a member of the Liberal Party. Eyde had bought Norwegian waterfalls to build hydroelectric plants, in cooperation with financial promoters in Sweden. In the well-known Rjukan waterfall in Telemark, which had a drop of 238 metres, he planned to build a power station. Birkeland had now taken out a patent for his invention. He believed that an electric arc plasma reactor could be produced in such a way that the connection between nitrogen and air could be exploited industrially in the production of saltpetre as an artificial fertilizer. That was the beginning of the cooperation on which the foundation of "*Norsk Hydro*", was created, originating from Kåfjord's hydropower station,

---

[140] Jago 2001:44

Northern Lights' research, and the beginning of activities on the summit of Haldde Mountain in Finnmark.

Norsk Hydro Research Factory, for artificial fertilizer production in Frognerkilen in Kristiania (Oslo). Photo: Unknown. Archive: Norwegian Museum of Science & Technology

It was discovered later that cooperation with Eyde was full of conflicts. Eyde wanted to have all of the honour for the Birkeland-Eyde-Furnace, even though it was completely clear there had been a division of work in connection with its production. Birkeland was responsible for the scientific work. Eyde was the entrepreneur, and Director of the Board for "Norsk Hydro", (Norwegian Hydro) from 1905.[141]

Some years later there was no doubt as to who should have the credit for the scientific basis for Norsk Hydro. In a showcase exhibition display at the Norwegian Industrial Workers' Museum, in Rjukan, we can read the following information.

> Birkeland developed the technology on which the "*Norsk Hydro*" industry was founded, the Arc furnace. The challenge then was to bind the nitrogen found in the air into nitric acid, which was then used as a source of nitrate, which could be used to produce artificial fertilizer. Birkeland's most important contribution in connection with this work was his research on controlling electrical arcs. To simplify this we can say that an Arc furnace, sends an alternating current through electrodes. In that way they form semi-circle light arcs. In the exchange between such arcs one can see a circular-rounded disc-shaped flame. In close proximity of the area with the highest temperature in the flame, nitrogen and oxygen are bound together and oxidization takes place.

It was not only Birkeland that had conflicts with Sam Eyde. In the same display we can also read that the change from industrial founder to industrial leader was difficult for Eyde:

> He was not a good administrator. He was very ambitious for himself and muddled together his own interests and the company's interests. Eyde gradually became a problem, as the Director of the Administrative Board of Norsk Hydro, and he left. Publicly it was presented as being a voluntary decision, but in reality he was given no choice.

In his time Eyde was highly regarded. When he became 60 years of age in 1926, the Norwegian newspaper, "*Aftenposten*", lauded him because he had, together with, Norsk Hydro, created a livelihood for twenty-five thousand people, where earlier only a thousand people had been able to live and work.[142]

In 1906 Birkeland sold his shares in Norsk Hydro, but later he had difficulty in getting the payment for the contract-agreed compensation for his contribution. The conflict with Eyde was long and bitter, and those working closely with Birkeland were very aware of this. Karl Devik, who worked as Birkeland's assistant throughout a number of different periods,

---

[141] The whole conflict is thoroughly accounted for in Egeland 1994:35

[142] Egeland1994: 14

later wrote a letter to his brother, Olaf Devik saying, "It would give me great pleasure, if at some point of time I could publicly tell how Minister Eyde, has acted to harm Birkeland in many different ways – "[143]

There is a wide spectre of known inventions produced by Birkeland. He had patented among other things: the electromagnetic cannon, hydrogenation of fat-production of butter, drying fish, processing of organic remains, a mechanical deaf-aid device and a hardening process for whale oil.[144] These inventions were, as mentioned earlier, motivated by his need for funds for his expensive research projects. It was not always easy to get consent for the grants he applied for. This is his comment:

> It can be claimed that thrift is a virtue that Norway as a nation quite certainly possess, maybe a virtue of necessity. But the nation's idealism often swings interchangeably towards a somewhat charming discrepancy regarding economy. The grant for my Northern Lights' expedition is an example of this.[145]

Some of the inventions were a result of unexpected discoveries. An interesting observation was made while experimenting with the earlier mentioned power switch that led to the discovery of the firearm known as, the electro-magnetic cannon. Birkeland suddenly discovered that a small piece of iron moved through coils at a speed of a bullet. From this observation he saw the potential of developing accelerating projectiles. That was the beginning of a weapon that could shoot at a target from a long distance. The patent for that discovery was his first. It is dated 16th September 1901. Simultaneously, "*Birkeland's, Firearms' Company*", was founded.[146] Naturally enough that discovery interested the governments in quite a number of countries, and had great potential as a source of income for him.

England was especially interested in the cannon, in connection with the beginning of World War I, in 1914. Despite the possibility of good earnings with the sale of the cannon for war service, Birkeland, who was in Egypt, took an initiative that did not ensure any economical profit. He sent letters to members of the, "*British Commission for the Examination of Inventions of War*". In his letters he offered the British Government free access to develop and use his electromagnetic cannon providing three conditions were kept:

143 Egeland1994: 34 (It is not clear what is meant by Eyde being called a minister.)
144 Egeland1994: 35-36
145 Devik 1971:3
146 Egeland 1994:24-25

1. Absolute discretion: Birkeland's name should never be used in any documents.
2. When a weapon was completed, it should also be given free disposition rights for Norway.
3. Weapons based on that technology should never be used against Scandinavian people.[147]

Birkeland never visited Haldde after 1910. He had obviously planned a visit. Dagny was expecting him. In her letter dated 23.1.1913 she writes that both Birkeland and Størmer were expected to come on a short visit in the summer. However Carl Størmer, who was also an acknowledged Northern Lights' researcher, and Birkeland ended up in a debate about Birkeland's Northern Lights' theory. Størmer supported other critical voices, some of them from English researchers, in crass assumptions written in the newspaper, "*Aftenposten*". Birkeland defended himself in the same newspaper against what he called, "a stupid attack", by Størmer. The argument continued in the newspaper with great intensity, until Birkeland ended with the following comment in the newspaper, "*Morgenbladet*" on Saturday 22nd November 1913. "All right. If Mr Størmer can get people to believe that he has ever had any big ideas in his Life, then that is so much the better, even though it is at my Expense."[148]

Dagny does not mention these controversies, but she writes to her mother-in-law that they had sent Birkeland a telegram from Haldde for Christmas, and this is how it began, "Birkeland, kind man, lots of money he does send..."[149]

---

[147] Jago 2001:224 ref. in Egeland & Burke 2005:106

[148] The academic contents of the debate between Birkeland and Størmer are written in Egeland & Burke 2001:98-100.

[149] Letter from Dagny to her mother-in-law, 23.0. 1913

# CHAPTER SIX

# A DANGEROUS EVENT

Snow stuck fast and coated the Observatory walls after an intense snowfall. Photo: Hilding Köhler Archive: World Heritage Rock Art Centre - Alta Museum

*Thank you for everything you have sent, it is wonderful. I can even use the egg warmers for the eggcups. We actually have two patients here now […]*[150]

Living on Haldde was not completely without problems. The assistants' lives are suddenly and unexpectedly in danger, only a month after Dagny and Ole have arrived. In the early morning hours of 30th October a distressed and dazed Hoel, their co-worker and 2nd assistant, woke the couple. Dagny describes the dramatic occasion in a letter to her mother-in-law the next day.

150 Letter from Dagny to her mother-in-law 31.10.1912

> Yesterday morning we were suddenly woken, and learnt that both assistants had been poisoned with carbon monoxide, fortunately they both woke up with a terrible headache, and Hoel was sufficiently awake to come running down to us.[151]

The letter does not actually say which assistant woke first, but it seems clear that Aabakken is mainly affected, as he is unable to walk on his own or with help from Hoel. Hoel is not quite as bad, and completely forgot that he could ring directly to the main house from the observatory, and that there was no need to run down, "He was probably unable to think clearly,"[152] Dagny remarks to her mother-in-law. It is possible that Hoel wanted to go down to fetch some medicine for Aabakken, not just to warn them what had happened? Dagny writes in her memoirs that Hoel, "wanted to fetch the first-aid box for medicine."[153] He was probably too ill to go up to the observatory again.

In her memoirs for her grandchildren Dagny describes what Ole did: "Grandfather got up just in his nightshirt, and found the medical books we had brought with us." Once he got up to Aabakken he thoroughly aired the observatory, and gave the patient what they had in the house. "We started with camphor drops and I had some milk."[154] Aabakken was so ill that he had to stay in the observatory all night. Ole also remains with him to see that all goes well, or as Dagny writes, "Ole has been acting as a nurse yesterday and today, and has been there all night.[155]

The assistants were very lucky that the headaches, which one gets from inhaling carbon monoxide, woke them. It is difficult to be aware of carbon monoxide gas, which has no colour, smell or any taste. The gas comes from living coal smouldering, and if one breathes in too much of it one can die. Those that sleep in air full of carbon monoxide will gradually become unconscious and then choke as they sleep, because of the lack of oxygen in their blood. The carbon monoxide in the observatory was because they burned coke slowly through the night, so that the room was warm in the morning.

[151] Letter from Dagny to her mother-in-law 31.10.1912
[152] Memoirs for grandchildren, written by Dagny, 1976
[153] Memoirs for grandchildren, written by Dagny,1976
[154] Letter from Dagny to her mother-in-law 31.10.1912
[155] Letter from Dagny to her mother-in-law 31.10.1912

Here is one of the assistants working with an instrument outside the observatory.
Photo: Hilding Köhler, Archive: World Heritage Rock Art Centre - Alta Museum

They use the same form of heating in the main house too, although Dagny and Ole have not had any problem with it. They discover that the ventilation in the observatory is not good enough. Later the residents of the main house experience that under certain weather conditions, they also have the same problem.

The piano is the central piece of furniture in the house, with its metronome and thermometer handy nearby. Photo: Hilding Köhler, Archive: World Heritage Rock Art Centre - Alta Museum

"Yes, so then it was, what shall we do?" Dagny writes.[156] She is uncertain if they have done enough for the assistants and would like to consult a doctor. She then finds a solution:

> I was sitting in bed when I suddenly realised that we had earlier had a telephone connection, because of weather warnings. The wire was still on the ground, as we did not have any poles to hang them on. We managed to contact Doctor Gjessing,[157] and he told us what we had to do.[158]

[156] Letter from Dagny to her mother-in-law 31.10.1912
[157] Full name: Rolv Ragnvaldssøn Gjessing
[158] Memoirs for grandchildren, written by Dagny, 1976

The day after this event the weather is good enough for Doctor Gjessing to go up with the drivers, to see the assistants for himself. The drivers actually went up the steep slopes on another errand. "Yesterday the piano came so we had a very busy day, as the drivers also needed something to eat, and the doctor stayed for dinner, "Dagny writes.[159] During his visit the doctor did not only check the patients, but also the observatory, and gave his opinion about what he thought was the reason for the accident:

> Frantzen was, according to the doctor a hard worker, but without any conscience, he had done a poor piece of work for the assistants living up there. He had put in an old door and completely neglected the whole roof. The doctor said it was against the law not to have ventilation in those rooms, and we can never have a window open there. Here in our house we have excellent ventilators, and in the bedroom the windows are open all day, unless there is too much of a storm.[160]

Dagny and Ole were aware that the condition of the two assistants was serious, but it is quite possible that they were not completely aware of the complications that could incur, that we know about today. Hoel was better within a day, but Aabakken took much longer to recover. "Aabakken was very ill, but Hoel has got up today. They have lost a lot of blood so they need nourishing food and green vegetables, and they have to take iron tablets for a while, as the blood corpuscles have broken down."[161]

The carbon monoxide the assistants have breathed in has combined with the haemoglobin in their blood, which in turn has prevented the intake of oxygen. The result of this was that their bodies did not get enough oxygen, which can then cause damage to different organs. Especially vulnerable are the larger organs, the brain and the heart. There are a number of long-term damages that can occur if there is a serious case of carbon monoxide poisoning, such as reduced cognitive ability or headaches, concentration problems and difficulty in sleeping.[162]

---

[159] Letter from Dagny to her mother-in-law 31.10.1912
[160] Letter from Dagny to her mother-in-law 31.10.1912
[161] Letter from Dagny to her mother-in-law 31.10.1912
[162] Norwegian Health Information (NHI)

The photos show the instruments used to gauge the electricity in the air. Photo. Hilding Köhler, Archive: The World Heritage Rock-Art Centre – Alta Museum, and The National Library of Norway, Photo collection.

The assistants show the typical symptoms of having been poisoned, headaches, nausea, giddiness, breathing problems and confusion.[163] It is doubtful whether the help they were given was of any use, except for the room being aired. On looking back later one can perhaps conclude that Hoel had got out into the fresh air and therefore recovered more quickly. Whereas Aabakken was lying in the room longer, and breathed in the gas, and therefore required a longer period in which to recover.

Dagny writes that, "Aabakken had little energy for a long time, and had to have extra food to help him get better".[164] Whether Aabakken had any permanent damage from this is impossible to say. The letters do not tell us much more about him, but Dagny has some small comments now and again. Among other things she writes to her mother-in-law about the challenges she meets in trying to satisfy those living there with regard to what they want to eat. She especially mentions the 1st assistant only 17 days after he nearly died, "Aabakken is the most difficult."[165] When a couple of weeks later after the accident one of his toes freezes, her comment is, "Ole never complains about things like that."[166]

Aabakken finally decides to leave Haldde, without his letter giving any clear reason for this. Maybe it was his health or the living conditions? Aabakken left in March 1913, five months after he was poisoned. Ole commented about Aabakken leaving to his mother:

> You were so sad about Aabakken leaving, but you do not need to worry any more about this. He was very nice in many ways, but on the other hand he was rather spoilt, so that things are somewhat better now, and more peaceful.[167]

Aabakken decides not to be 1st assistant for the research station's pioneering work. A reorganisation must be made and Hoel takes over Aabakken's job for a while. "Hoel would very much like to be 1st assistant, and I think I will try and keep him in that job, but it has to be advertised, although I believe he will get it. He is at present placed in the position of acting 1st assistant, until further notice."[168]

The serious incident could quite easily have cost the lives of both assistants. Despite the shock the accident has given them, Dagny feels it is

[163] Norwegian Health Information (NHI)

[164] Memoirs for grandchildren, written by Dagny, 1976

[165] Letter from Dagny to her mother-in-law 17.11.1912

[166] Letter from Dagny to her mother-in-law 10.11.1912

[167] Letter from Ole to his mother 13.03.1913

[168] Letter from Ole to his mother 17.03.1913

important to write to her mother-in-law the following day, about more cheerful things. "It is a lovely sunny day today without any storm, but I do not think it will last for long. You probably realize how happy we are about getting the piano into the house, even though there were a number of problems getting it up here."[169] The happiness of having a sunny day after the storm is perhaps increased by the fact that the piano they had longed for was now in place.

[169] Letter from Dagny to her mother-in-law 17.11.1912

# CHAPTER SEVEN

# HOUSEKEEPING AT A HIGH ALTITUDE

Dagny enjoys a windless and sunny day on the top of the mountain. Photo: Ole Andreas Krogness. Archive: National Library of Norway - Photo collection.

*In general when it is just us that buys something, we have to pay double the price, which does not make one feel very friendly towards the shopkeeper.*[170]

As Dagny is married to the Director of the observatory, she is also responsible for the housekeeping for all of those who live there. There are five adults, who live on the top of the mountain, and her work consists of making the food, cleaning the house and washing clothes, as well as some administrative jobs, such as ordering wares and food for the household, and keeping the accounts. She is responsible for organizing and planning

[170] Dagny's diary from 1912

the household's requirements a long time in advance. She has to find out what she can manage to buy, estimate how long it will take to transport, and not least of all ensure that she gets a good price, that will provide cheap and nourishing food. Altogether a much more demanding job on Haldde than she had thought when she was down south, "My housework is more than I had calculated, […]".[171] In her memoirs she writes:

> It was a big step to come to live on the top of a mountain, more than 900 metres above sea level […]. I was responsible for the food, but there was usually a maid to help me. The first year we had both assistants and a foreman to provide food for. I had not learnt how to make food before I got married, so I used a cookery book and read how food ought to be cooked, and most of the time that went quite well. But it was not always easy. We did not have any milk or water. We had to melt the snow both for food and washing, and that was hard work.[172]

## Advanced Cooking for Beginners

Dagny has always been used to a maid or housekeeper making the meals, from the time she was a child, and at the student home the meals were provided. On Haldde she has to prepare food for other people for the first time. Her inexperience means that she has to start from the bottom and learn the basic principles, while at the same time she has to provide meals for the other residents. That must have been challenging for her, and Dagny says in her letters that she would have been happy not to have to prepare food for others. She wants to get out of the arrangement, but realizes that it can be difficult.

> It would have been lovely to be alone and not to have the two men, but they were so pleased not to have to prepare meals and it was quite a lot cheaper for them to eat with us and not have to have a maid. […] I have thought that it would be cheaper for us to cancel the arrangement with the assistants, and then have a maid to ourselves, and see if I can manage with that. […][173]

Dagny realizes that to tell the assistants that they have to provide their own food was not possible, and she would just have to accept the situation. That is why such an arrangement continued for the first year, but

[171] Dagny's diary from 1912
[172] Memoirs for grandchildren written by Dagny 1976
[173] Letter from Dagny to her mother-in-law10.11.1912

the following year the assistants began to make their own food, something that Dagny is looking forward to.

Despite not having any previous knowledge Dagny just gets on with doing what is expected of her. She makes a strategic plan and discovers that Hanna Winsnes' cookery book is in fact her salvation. Hanna Winsnes was a pioneer within cookery, and known to have contributed to making domestic science work for housewives a professional occupation, in the latter part of the 19th century. Her textbook, "*Lærebog i de forskjellige Grene af Husholdningen*", published in 1845, is about different branches of housekeeping. It contained many good recipes for a housewife, and gave instructions about how one prepared food from the butchering stage up to more everyday tasks. Winsnes also stressed the importance of cleanliness and utilizing food provisions. The cookery book Dagny uses is a reprint of the first publication,[174] and in 1912 a popular and frequently used book. It is a book that had been constantly published in new versions, the last one as late as 2004.

The residents could enjoy the fantastic view from the front door, when the weather allowed it. Archive: World Heritage Rock Art Centre - Alta Museum

Winsnes does not only publish food recipes, but also gives information about what tasks a wife is responsible for when managing a household, something that gives us an idea about what kind of role is expected of Dagny. Amongst other things Winsnes criticizes the new educational methods of that time, which did not give girls enough experience for them to be able to tackle such a task properly.

---

[174] Hareide 2009

> I am completely convinced that it is every young wife's and engaged girl's good intention to manage their housekeeping to the advantage and satisfaction of their husband, by doing things in the way they think they should. Unfortunately, the newer educational methods do not make it possible for most young ladies to have Domestic Science lessons, and the inexperienced move forward uncertainly into the unknown reality, which results in slow hard work, and therefore many lose interest and stop at the beginner stage. Even so I will give them some assured consolation that any kind of work becomes interesting and easier with practice and attention. It would please me that what I write could give some small support to hold onto, on their difficult journey, and I really hope that they, with pleasure and glory, will reach their noble goal.[175]

Winsnes' description is in many ways similar to that of Dagny's situation. She is a young wife, who during the years when she was growing up had not had enough experience about how to manage a home for her husband, so that he could do his work and be satisfied. In Kristiania she could have asked for advice from her mother or mother-in-law. On the top of a mountain it is Mrs. Winsnes in the form of a book, which is her support. She reads it when she goes to bed at night, and during the daytime she tries out the suggestions.[176]

It is a cumbersome and demanding housekeeping Dagny is responsible for. She works as hard as she can to develop order and routines.

> The whole morning is mainly used to prepare dinner. After I have made our bed and tidied up a bit, I go out to my meat and fish boxes and find some food, if I have not taken it in to thaw the previous day.[177]

Dagny uses the snow as a storage place for the food, and hermetically preserves the rest of this food when it begins to thaw in the spring. "To begin with we had to have the fish and meat buried under the snow, because we had no building in which we could put it. It was rather difficult having to dig up dinner every day."[178] Unfortunately not everything goes as planned. When they get a dog, they discover dog stools when the snow melts, not very far away from where they kept their food. They have no other choice than to get rid of the dog. One of the

---

[175] Winsnes 1845:1

[176] Memoirs for grandchildren written by Dagny, 1976

[177] Dagny's diary from1912

[178] Memoirs for grandchildren written by Dagny, 1976

carpenters, who works in the house, takes it home to the delight of his children,[179]

Dagny is not completely alone with the housework. She has a maid, who she hopes will help her in both the house and the kitchen. One of the problems with this is that the girl is very young, and knows very little about how to cook food. Once Dagny has learnt how to make something she has to teach the maid, who is then expected to be able to help her. In December, when she begins to feel that the work is getting a bit more difficult, because of the child she is expecting, she writes to her mother-in-law. "I am now going to begin to teach the girl how to make food and I have to tell her why she will be responsible, otherwise I will have to stand there and make the food myself.[180]

That Dagny has someone to take over when she gets closer to the end of her pregnancy is important. She feels that her double role as cook and teacher can at times be difficult, "It is really hard to have a maid who cannot cook, especially when we have visitors, but I am sure she will improve."[181]

## The Food Budget Causes Stress

Most of the information about the purchasing of food that they need during the first year, is written about in letters. As well as cooking and planning meals, the actual ordering and buying of provisions is new for Dagny. She confides in her mother-in-law, "You say that I must not completely spoil them with the food I give them, but that we must be able to pay what it costs."[182] Those living there have different requirements, which Dagny tries to consider when she buys the provisions. It worries Dagny that the food will become too expensive, as well as it being more complicated to make.

---

[179] Oral source: Gunvor Jensen, Toril's deceased aunt. The carpenter was Toril's grandfather Henrik Jensen. Both he and Toril's uncle, Isak Sørstrøm, can be found in the list of workers on Haldde. See Nielsen & Peterson (1993) Nordlyspionerene

[180] Letter from Dagny to her mother-in-law 15.12.1912

[181] Letter from Dagny to her mother-in-law 08.01.1913

[182] Letter from Dagny to her mother-in-law08.01.1913

The snow that is to be melted for water, had to be carried from further away when there was mild weather, something which meant that Ole Andreas Krogness had a number of practical tasks, in addition to his research work. Archive: World Heritage Rock Art Centre - Alta Museum

> They are so different, […] there are some things Ole does not like, and that the others do, for example herring, and I feel there should also be some meat or something else on the table, as he does not eat cheese. Now and again I warm up the rest of the meat and fish if I have some, because then less is eaten of the other things.

Keeping a balance between making meals that everyone likes, being economical, and at the same time not giving differential treatment is difficult.

Dagny becomes more experienced, and she develops good routines that utilize her food provisions in a better way. In one of her letters she describes how she provides good nourishment, as well as making some economical profit.

> Having boarders in our home is not that easy, and as they pay for their food they must have nourishing food. Today it is meat soup, which is

> cooked from a bone at least three times. I vary it by adding different things, which they like very much.[183]

Hanna Winsnes, Dagny's chosen teacher, recommends, "A good housewife must buy in provisions for the home when they are at their cheapest, which is different for different things, while simultaneously keeping a detailed account of the expenses."[184] Dagny uses her diary to keep her accounts during the first part of their stay. One way of keeping expenses down is to buy provisions from a shop that has the cheapest delivery service, which is something Dagny struggles to do. In Kåfjord there is only one shopkeeper, Heitmann, so she does not have much choice. It also seems that she is afraid of being cheated by him. Dagny poured out her frustration to her mother-in-law:

> They are terrible about putting an extra price on everything; just imagine they weigh everything that they have to deliver, and add something extra per kilo for every single thing, so that Heitmann can get transport costs. This means that food is 5 øre more expensive per kilo. There is no advantage in provisions being cheaper here. I am extremely anxious that I am making far too expensive meals.[185]

Dagny thinks that her money is used up far too quickly, especially as Heitmann takes an extra price per kilo for transport expenses. Her irritation is clearly shown in her letters. That they are charged extra for the transport of goods that one of Heitmann's staff takes up to them is too much. For example, Dagny has to pay twice as much for the potatoes that are delivered up to Haldde. Together with the extra weight charged and transport it no longer costs 6 øre, but 11 øre.

Later in the year Ole solves the problem of transport costs by buying a reindeer, and paying a driver, so that they are no longer dependent on Heitmann. Later the annual accounts for the observatory show that they pay for their own horse-drawn transport, for the next few years. They experience quite an amount of trial and error during their first year, but Dagny does not give up, and always tries to find new solutions to save as much as possible.

> I have had a baking trough and a spade made, and I have now baked the first bread. It was very much cheaper than Heitmann's as he charges 25 øre

---

[183] Letter from Dagny to her mother-in-law10.10.1912

[184] Winsnes 1845:48

[185] Letter from Dagny to her mother-in-law, 19.10.1912.

for his bread. This month we have made just as much bread, but we have only used 5-6 kroner now, which is the cost of a sack of flour.[186]

It is very steep up to the small colony. If the ice is too thick on the cable, and stopped the aerial tramway *(kibben)*, there were quite a few hills to climb with the goods. Archive: World Heritage Rock Art Centre - Alta Museum

However, it is difficult to save money by baking bread when the yeast freezes and the bread does not rise, "I am very unfortunate with my bread making, but will persevere until I succeed. Usually the yeast we get is frozen hard, and I think that is the cause."[187]

Dagny is also aware that they can save money in other areas, for example by doing the washing themselves. Now they send dirty washing down to the village and pay 12 kroner each time,[188] Dagny does not dare to think about what will happen when she gives birth, as then there will be more washing. "If only I had an attic where I could dry our washing then we could melt snow to wash the clothes, but the attic is being used […] so things hang to dry in the dining room and kitchen."[189]

---

[186] Letter from Dagny to her mother-in-law, 10.11.1912.
[187] Letter from Dagny to her mother-in-law, 23.01.1913.
[188] Statistics Norway.
[189] Letter from Dagny to her mother-in-law, 17.03.1912.

The question is whether Dagny was preoccupied with saving money, because she is afraid of using too much, or because it is more expensive to live on Haldde than elsewhere? Devik says in an interview some years later, "No we did not experience any need, both Krogness and I had a professor wage about 4.000 kroner (ca. 218.105,00 kroner in 2016[38]) a year, we could easily manage on that."[190] When a working family during the years 1912-1913 had to survive on 1.986 Kroner (ca. 108.289,00 kroner in 2016[38]) a year for a whole family,[191] perhaps we can assume that it had not been necessary for Dagny to have been so thrifty.

## Worries and Frustrations

Post and provisions are driven up to them three times a week. It is apparent that Dagny is worried about the three litre of milk they order every other day. There is, of course, also hermetically sterilized milk, "but fresh milk is the best", she writes. Her uneasiness is due to her own thoughts, "that their milk probably came from cows that had tuberculosis, because some of the people in Kåfjord did not have healthy cows."[192] If the milk had not been treated in the right way, then everyone can be infected and be seriously ill. In 1912 the pasteurizing of milk has not begun, and therefore milk could easily be a source of infection. At that time it was normal to boil the milk before it was used, to be on the safe side. Despite doing this Dagny was still worried.

Tuberculosis was a terrible illness. In Kristiania, Theodor Frølich, carried out an examination of 3000 children from 1912 to 1915. The result was disheartening. 90 percent of the working-class children and 80 per cent of the total population were infected by the time they started school.[193] Finnmark was hardest hit by tuberculosis, and was clearly worse off than the rest of the country. Poverty was the usual explanation for being contaminated, as well as the bad living conditions.[194] The regional medical officer, Andreas Bredal Wessel, in Sør Varanger, in Finnmark had this to say:

> It was the inhabitants' lack of resistance, due to their poverty, bad living conditions resulting in uncleanliness, poor hygiene and ignorance, as well

---

[190] Cutting from," *Uniternt"*, unknown date. "It was good to live on Hal'di, and meet Norwegian researchers' grand old man": Olaf Devik.
[191] Wig 2007:155.
[192] Letter from Dagny to her mother-in-law, 19.10.1912.
[193] Wiig 2007:154.
[194] Elstad & Hamran 2006: 158, 322.

as a lack of knowledge, which basically made them both susceptible to the infection, and hampered the possibility of fighting it.[195]

For Dagny it was important to get her milk from healthy cows in Kåfjord. Archive: World Heritage Rock Art Centre - Alta Museum

A vaccine against tuberculosis had not been discovered, and there was no medicine that could be used to combat it. It was very infectious, and the only treatment was isolation and rest. As Dagny is in charge of the housekeeping, she is responsible if anything goes wrong, so she tells her mother-in-law that she has talked to Heitmann to make certain that the milk is not infected.

Even though Dagny regards Heitmann as a resource person in certain areas, her letters show that she thinks the shopkeeper sells them bad quality wares. In the late autumn Dagny solves this problem by ordering food provisions from Trondheim or Tromsø, and arranged for them to be transported to Kåfjord. She writes the following about the shopkeeper in Trondheim, "He has good wares, because Heitmann has rubbish, there is hardly a thing that he sends that is not without fault".[196] Dagny must now calculate a longer length of time before her orders arrive; in fact, a whole three weeks from when she first sends her order, until the provisions get to Haldde. She plans her orders for the household's daily requirements, and

---

[195] Wessel A.B. Tuberculosis in, "*Sydvaranger*", doctor's district. 1886-1911. Extracts from the medical report 1911. See Forsdahl 2002:2673.

[196] Letter from Dagny to her mother-in-law, 10.11.1912.

for social occasions extremely carefully. Nevertheless, sometimes the food does not arrive at the expected time. This is mainly due to the bad weather along the northern coast.[197]

Dagny works hard to provide good quality provisions so that those living on Haldde can have varied and nutritious meals. Information about the right kind of food was an important part of the work done to improve people's health in Norway from 1910, and up to the Second World War. As the daughter of a health-conscious doctor, Dagny is up-to-date with this situation. Probably better-off families in Finnmark followed the health advice that was given, but for the majority of the population in Finnmark it was not a very realistic situation. They had enough to do looking after themselves, and could absolutely not afford to order goods from Trondheim, like Dagny and others like her. The diet of people in the region at that time was generally bad, which to some extent was due to a shortage of potatoes, other vegetables and fruit. The farmers in the county did not cultivate enough to cover the district's requirements, and deficiency diseases were spreading.[198] There was great poverty. Those living on Haldde had other conditions, and with Dagny's diets there was no danger of deficiency diseases. Her worry was to keep within the budget, but it was a budget that could cope with the price of both raw wares and transport.

As Dagny and Ole settle into their new environment, the letters change and paint another, and maybe more nuanced picture, about her experiences with local products. Dagny now understands the reason for the poor quality of the food, and that it was not necessarily Heitman's fault, but was more likely due to the fact that people in Finnmark got bad products from the south of Norway. She became aware of the connection between local frustration and the general situation in Finnmark:

> I think I will get my provisions from Thj, (Trondheim), as Heitmann does not have such good wares. The tablets for example melt straight away in your mouth. Although, apparently it is not Heitmann's fault, because I heard the doctor say that they send bad wares to Finnmark.[199]

Despite worries and frustrations it seems that things begin to fall into place during the first year. Dagny writes to her mother-in-law that the maid is much more capable in the kitchen, "The maid has become a clever

---

[197] Beichman 1939:16.

[198] Bore 2007, Beichman 1939: 145, Forsdahl 2000.

[199] Letter from Dagny to her mother-in-law, 31.10 1012.

cook now, so I feel less worried as far as that is concerned. But, good heavens, such a lot of money has been used."[200]

It was not unusual for the residents to have to dig themselves out after a hard snowfall. Here the entrance to the kitchen is nearly blown in by the snow. Archive: World Heritage Rock Art Centre - Alta Museum

Dagny also discovers that ordering provisions at different times of the year is demanding, and she writes about the choices she makes, rather than about her frustrations.

> I still have some French beans, although they will not last long, but they can be frozen and are just as good. If I have to buy some I have to have them chopped and salted, but that is quite expensive. Root vegetables I bury in sand and they keep all winter. The potatoes I get from Elvebakken, as they are good, but we are thinking of having them in the machine house, because it does not freeze in there.[201]

[200] Letter from Dagny to her mother-in-law, 13.03.1013.
[201] Letter from Dagny to her mother-in-law 03.09.1913.

# CHAPTER EIGHT

# PROFESSIONAL WORK AND DAILY LIFE

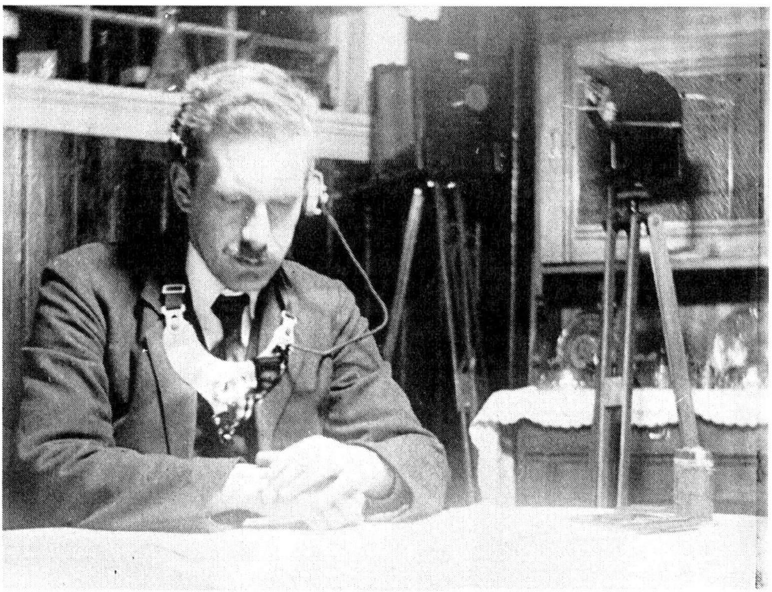

Ole in a concentrated telephone conversation at the dining table. Archive: The World Heritage Rock-Art Centre – Alta Museum

*I am not even allowed to touch the desk. If anything is missing, it is immediately said that it is probably the wife who has been tidying up.*[202]

In most letters from Dagny she mentions the research that Ole is doing, perceived from her point of view. As soon as they arrived on Haldde she started to write a diary, and writes about the magnetic instruments that:

[202] Letter from Dagny to her mother-in-law, 29-11-1912.

> …require constant care. They are underground in a square room. The exterior and interior of the room is covered in pitch to make it waterproof. That is to make sure the instruments fluctuate regularly (at least that is what I understand). They have a bell that rings every hour, to the great annoyance of the maid, who thinks it is for her.[203]

The cable by the brook where the steepest scree up to Haldde starts. Archive: National Archives of Norway

We have a phone too, she writes:

> It is constantly ringing during "Northern Lights' evenings", as we call them. It is namely a Dr. Vegard from Bossekop, who takes photographs of the Northern Lights in the evening, and he communicates with Haldde every evening from half past seven.

It was a huge programme of work that Ole would implement on Haldde. It included observations and registrations of all meteorological elements, such as temperature, humidity, pressure, wind, rainfall, weather and clouds. To understand the electrical currents in space, he also had to observe and record conditions that had to do with terrestrial magnetism, such as atmospheric electricity and earth currents. In addition, he had to observe and photograph the Northern Lights, and link these registrations.

[203] Dagny's diary from 1912.

There were many notes and much deskwork to do, when he was not busy with direct observations.[204]

It was not easy for Dagny that the workplace is actually also their home, and that working life and home life overlapped. Right at the beginning of their stay they took a photograph of Ole at his desk, which Dagny said was in a shambles. She did not want this to be photographed, as she knew that the housewife is blamed if it is untidy in the house.

Probably the couple had an agreement as to how their work should be organized practically, giving Dagny the task of reporting to Ole's mother, because Ole had so much work to do.

A work place in the home. Photo: Ole Andreas Krogness. Archive: National Library of Norway - Photo collection.

In her letters Dagny included some lines about the research, but sometimes Ole himself had to report to his mother. The work could

[204] Evjen 1945:15.

interfere both with Dagny's domestic responsibilities, and the time Ole could find to write letters. On 10$^{th}$ November 1912, she wrote in a slightly annoyed tone:

> Yes, I am happy Vegard is not here because he is such a fusspot. Yesterday they kept on taking photographs of the Northern Lights until late, so we were not finished with supper before 10.30 p.m. […] this evening it is also the Northern Lights. Vegard has at least called us six times during a quarter of an hour, which meant that Ole only just managed to get to the headline on a letter.

At this point of time Dagny did not know that Vegard's Northern Lights' research, would later be acknowledged on an equal footing with Birkeland's. For her he was mostly an annoyance in her daily routine. Two to three weeks later, she experiences, on several occasions, how her domain becomes invaded by research activities:

> Today, all of a sudden, Ole became a cleaning maniac here in the dining-room. They pulled in all the bowls containing fixer solutions, so all the corners of the room are filled with bath-tubs. I have given up on cleaning until we get organized. To begin with I was so eager, but now I am just waiting patiently.[205]

It soon became clear to them both that coping with practical matters on the top of a mountain was demanding, and time consuming. Ole tried to make everything easy for their domestic life, and Dagny thought he was magnificent. He had no one to help him. He even melted snow in a brewer pan so that they had water.

[205] Letter from Dagny to her mother-in-law, 20.11.19112.

Measurements of the Northern Lights had to take place outdoors, here on the top of the observatory. Archive: World Heritage Rock Art Centre - Alta Museum

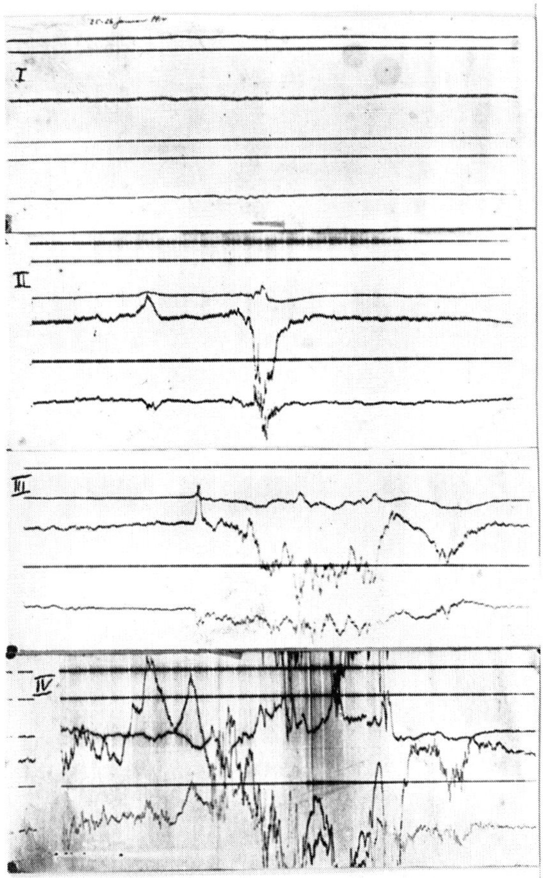
I
II
III
IV

NORDLYSJOURNAL
Det Magnetiske Byrå
40

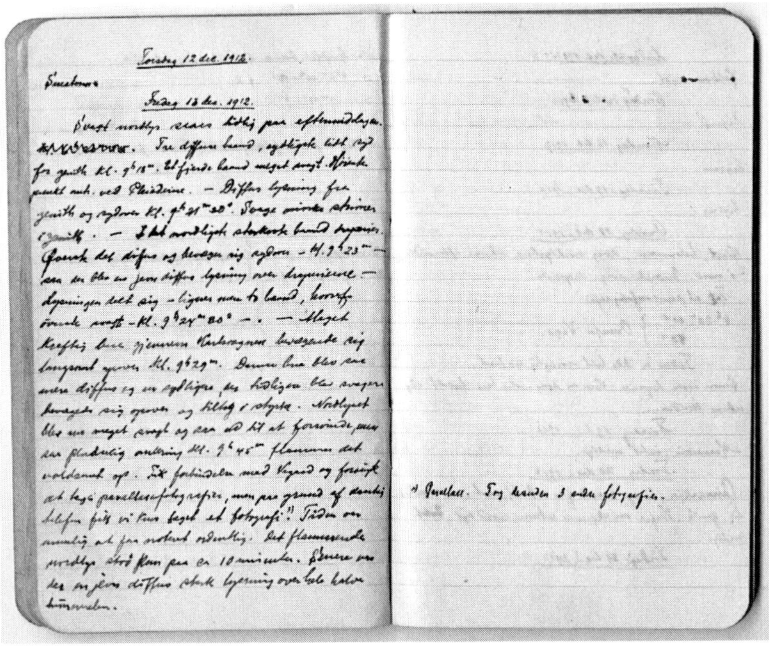

Registrations and records that show some of Ole's research work. Archive: Haldde Observatory in the Regional State Archives in Tromsø.

He also chopped wood, and in fact such physical work was good for him, in Dagny's opinion. At the same time she was a little worried that such work would be too much for him when he seriously starts writing. Ole writes to his mother:

> Things are reasonably in order up here now. We have installed ourselves in all three rooms, and we have a really good attic combined with a laboratory for terrestrial currents, and a carpenter's and instrument maker's workshop, in addition to housing for the workers. Observations are also well underway, and we have observed some quite interesting things.[206]

In continuation to this however, Dagny says that all the practical work could in fact be detrimental to the research work that her husband should be working with:

[206] Letter from Ole to his mother, 18.12.1912.

> Ole is struggling with all other kinds of work rather than mathematics and calculations. One minute he is in the engine room, and the next minute he is clearing away the snow. Then he has to take another day to do his accounts, which are very complex and complicated.[207]

However, Ole succeeded in writing an article for the newspaper, "*Aftenposten*", in January 1913. Dagny comments:

> Now Ole has written a long article for "*Aftenposten*", and hopefully all of you will enjoy it. He has worked day and night to complete it. He has so much to do, just with arranging and looking after everything, besides emptying the toilet, melt the snow and lots of other tasks. I do not understand how he has any time for research, and he writes in English, too. There is such a lot of correspondence. He has 16 letters at a time to answer, plus all the unanswered ones.[208]

All their correspondence was done by hand. It is not difficult to imagine that writing letters must have taken a lot of time, both those connected to Ole's scientific work, and those connected to administrative and private matters. It was through letters that they informed their network and maintained their connections, not only with their family, but also with the government, as well as with Birkeland, and with other collaborative partners in general.

Dagny was concerned about the impending birth of their first child, and tried to discuss, with Ole, what they should do in connection with this. Evidently it was not so easy because, "He has so much to think about, poor Ole [...]."[209] Ole commented to his mother about Dagny's condition in a letter, "Dagny is growing round and plump but flourishing. Occasionally she has some stomach ache here and there, but thankfully this passes quickly."[210] Dagny somewhat disagreed and continued in the same letter, "I see from Ole's letter that I am very plump, but it is not that bad."

---

207 Letter from Dagny to her mother-in-law, 08.12.1912.
208 Letter from Dagny to her mother-in-law, 23.01.1913.
209 Letter from Dagny to her mother-in-law, 23.01 1913.
210 Letter from Ole to his mother, 10.03.1913.

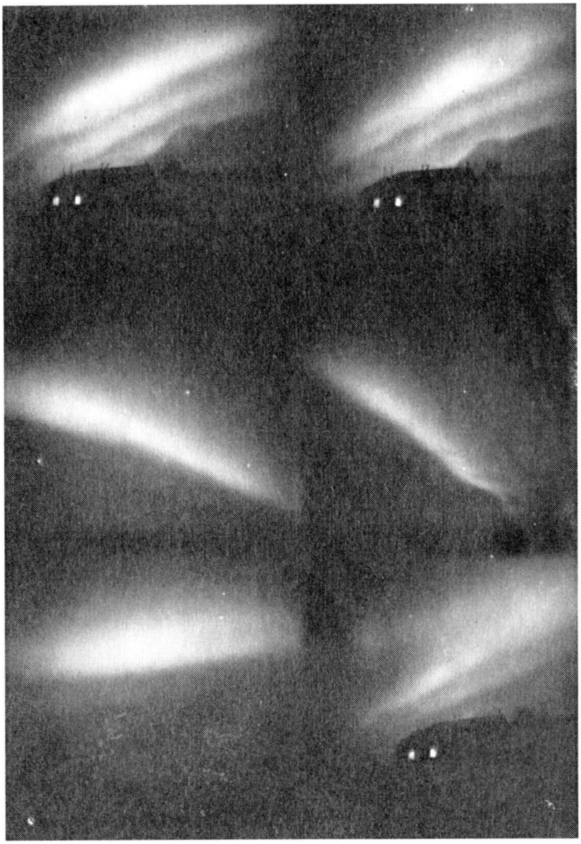

A photo taken at night of the house on Haldde, with Ole's inscription "Fig. 29, 10th January 1913". Archive: Haldde Observatory in the Regional State Archives in Tromsø.

Later in the spring, she could tell her mother-in-law that, "Ole has predicted the weather very well for the last fortnight, and has also made a lot of new discoveries that he is extremely happy about, but unfortunately I do not understand them."[211]

Ole was now well underway with his Aurora photography. At the same time, he took some photos of Dagny busy with her activities. In a letter to his mother he attached one photograph.

[211] Letter from Dagny to her mother-in-law, 13.03.1913.

> I am among other things taking cinematographs of the Aurora and other things. Here you see Dagny, who is searching for something in the food boxes. I also made an "Aurora cinematograph" which works excellently, but it could have been a bit more exactly constructed here and there. It consists of some planks, which are nailed together in a reasonable manner, and with this I can take 60 photos on each of the cassettes. I have taken between 100 and 200 photos the last two evenings. I will wait and see whether I get good results.[212]

The letters at our disposal are valuable research material. Nevertheless, there is quite a lot, which is not being said. In the silence between the lines there may be some possible information that we have no access to. For example, there is little information about Birkeland in the letters that went between the couple, and Ole's mother, despite the fact that the Professor was continually of great importance to Ole in his work on Haldde.

While both Dagny and Ole stayed in Bossekop waiting for the birth, they passed the time nurturing contact with acquaintances in the community. Among them was a person called Størmer. One day Dagny writes to her mother-in-law:

> We have been visiting Størmer and looked at his pictures and photographs. Ole seemed to be rather anxious about Mohn. Birkeland, he –Ole- is keeping silent about it as long as possible. Yes, I wonder if we will get any money to do some building? I now have my own plans for this.[213]

The letter shows that they have already decided to apply for an extension to the house to get more space, and Dagny has plans for how the house should be designed.

She makes no mention of what excited Ole about Mohn. This is March 1913, and we know that Birkeland left Norway that year for a longer stay in Egypt. His conflict with Carl Størmer is famous through newspaper articles, in which they are more or less yelling at each other. Størmer is in Alta taking photos of the Northern Lights. The question is, how does Mohn fit into this picture?

Professor Henrik Mohn (1835-1916) had founded meteorology as a science in Norway. He was a manager of the Meteorological Institute, where he organized weather observations that were to be the foundation of the world's first weather forecast service.[214]

---

[212] Letter from Dagny to her mother-in-law, 17.03.1913.
[213] Letter from Dagny to her mother-in-law, 21.03.913.
[214] Egeland & Burke 2005:159.

Henrik and his wife Julie Mohn had been neighbours with the family Birkeland in "*Nordahl Bruns gate 8*", in Kristiania, and they had encouraged their neighbour's son Kristian to follow his desire for a career in the natural sciences. They knew each other very well. Sometimes the boy was allowed to go into the office of the professor, where he observed different instruments and forms that Mohn used in his research.[215]

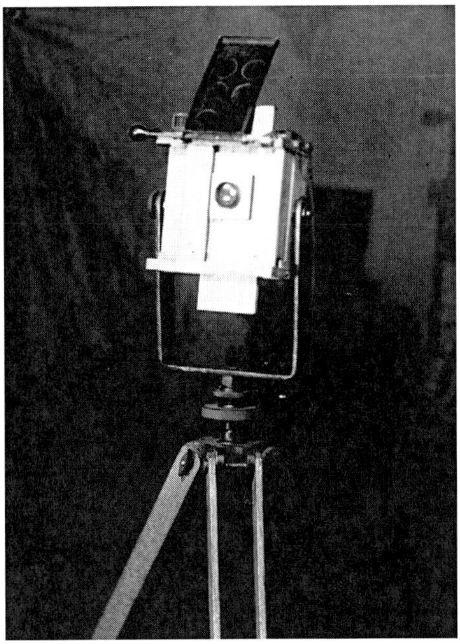

This is the camera that Ole built, called the "Krogness-camera". Photo: Ole Andreas Krogness. Archive: National Library of Norway - Photo collection.

It was through his connection with the Mohn family that Birkeland met their niece, Ida Charlotte Hammer, whom he later married and then was divorced from in 1913, before he travelled to Egypt. Mohn's wife, Julie, was born and raised in Alta, so there is a possibility that Henrik Mohn was the one that Dagny said Ole was inquisitive about. In her letter there is an incomplete sentence that Ole is keeping silent about something, but what, she does not say. It is clearly something connected to Birkeland, but she hardly mentions his name. Mohn had, as we know, had objections to

[215] Jago 2001:71.

Birkeland's plan from 1910 to establish a permanent observatory on Haldde. He wanted it to be built in Bossekop. In a letter to Sæland in 1910 Birkeland made an acid comment about Mohn's desire to place the observatory in Bossekop, "Mohn is married in Bossekop, and he thinks that the name Bossekop is so famous that the observatory should be placed there."[216]

It is tempting to think that some of Størmer's conflict with Birkeland was discussed during their visit. Perhaps Mohn made comments, not only about these professional disagreements, but also about the divorce between his niece Ida and Birkeland. Ole's reluctance to discuss controversies of all kinds may perhaps be the main reason why he would "be silent as long as possible", regardless of what might be said on that matter. Dagny loyally follows her husband's reluctance concerning conflicting issues, but she is unable to resist the temptation to infer a little bit about it in her letter to her dear mother-in-law. It seems as if the mother-in-law is well informed about what was going on, within the academic circles surrounding her son.

In a previous letter Dagny had remarked with a sigh about writing letters, "Ole thinks I should just be telling about happy occasions, but that would be unnatural for me."[217] It was not just the bright side of their existence that was important to Dagny.

[216] Letter from Birkeland to Sæland, 1910, National Library, letter 144.
[217] Letter from Dagny to her mother-in-law, 19.11.1912.

# Chapter Nine

# Dagny's Struggle

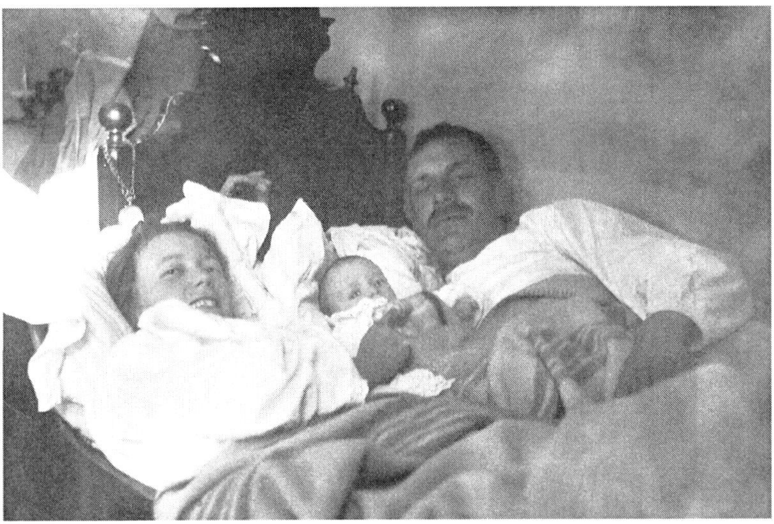

The young family a short time after the birth in Bossekop. Dagny, baby Oa and Ole. Archive: National Library of Norway - Photo collection.

*I already had to leave at the beginning of April, because the road down could be difficult.*[218]

The first winter on Haldde was busy, and made great demands on Dagny. She has for the first time in her life set up house, a research station, and home, where daily life and research would live side by side. She has tried to do the housekeeping in the best possible way, so that the situation for research work is satisfactory for Ole. Now other things are bothering her. She will be giving birth to her first child far away from her family and people she knows, and here she sits isolated on the top of a

[218] Letter from Dagny to her mother-in-law, 23.01.1913.

mountain, with just her husband and a maid to help her. Where is she to give birth? Should the birth be there, on the top of Haldde Mountain? It was a difficult decision.

Dagny seldom writes precisely about her pregnancy, and little about how she feels or thinks. It is mainly the practical events that she writes about to her mother-in-law, for example about buying provisions, knitting and sewing. Dagny is pleased that her mother-in-law gives her a helping hand, as she is good at sewing, whereas Dagny says that she is inexperienced. "Yes, many thanks for everything you do for me, my mother says you have worked very hard with the clothes, but you can also send them up to me, so that I can do some sewing."[219] The letters document that Dagny's mother also helped, and that they exchanged letters with each other, even though none of these letters have been kept for posterity.

Dagny has begun to be concerned about living on a mountain, and being able to get the things she will need when the baby arrives. She continually discovers new equipment that will be necessary. In her letter to her mother-in-law she orders new supplies from down south. "Must buy an extra bed. It would be a good idea to get some more nappies, in case we have no water, and I have to throw them away."

She prepares the baby's cradle, "I have collected some grouse feathers, but not that many, although it is written that it is best for the baby to lie on something hard."[220] She wants to have the cradle on a box, so that it is a bit higher, then she will not have to bend down so much. She does not want runners to rock the cradle, as she writes in the same letter that they are harmful, "according to modern-day advice."

There is no information in her letters about how her pregnancy is progressing, but as it is the first time she is to give birth, it must have concerned her. In just a few of the letters Dagny does say something more directly about the baby that is growing inside her. She calls it the "Twig", maybe it is an illustration of a new branch starting to grow on the family tree. "I am knitting socks and woollen vests for the Twig, it is also fun to practice…."[221]

Dagny uses Doctor Greve's book,[222] as a source of information and advice about pregnancy. Dagny feels uncomfortable discussing, "her circumstances", with the local doctor. According to the letter she writes to her mother-in-law, she and Ole get a chance to talk to the doctor about the

---

[219] Letter from Dagny to her mother-in-law, 25.01.1913.
[220] Letter from Dagny to her mother-in-law, 25.01.1913.
[221] Letter from Dagny to her mother-in-law, 23.02.1913.
[222] Greve 1906.

coming event, when the doctor visited their assistants earlier, in connection with the carbon monoxide poisoning, "We were able to ask the doctor about quite a few things, that is Ole did as I was too shy."[223] Dagny is four months pregnant then, and is aware of various changes in her body. The letters do not tell us what they talk about, only a reference that a conversation took place with the doctor.

Most people would seek advice from their close family and friends, when making difficult decisions. For Dagny it is only possible by telephone, and only from Alta, otherwise she has to write letters. Ole is the only one she can talk to. To confide in the young maid is out of the question, and she also feels that it is difficult to just chat with the assistants. At times Dagny struggles with loneliness. She longs for someone she can talk to, but she has very little hope that she will meet anyone cultured enough for her to discuss her problems with; while living high up on the summit of a mountain. When she writes that they will be getting a new assistant when Aabakken leaves, she has this to say:

> As far as I am concerned I dread having an uncultivated one when we have to be together so much, and with a bad smell in the rooms, ha-ha! Yes you think that I am perhaps too genteel now, but it may be the Twig that wants fresh air.[224]

Dagny does not regard either the maid or the assistant as good discussion partners.

The letters do not say anything about an examination, while she is pregnant, by a doctor or midwife. In 1913 Finnmark had no antenatal clinics for pregnant women. These were first founded in the 20th century. In 1920 it was just Bergen, Trondheim and Oslo that had such clinics. In the rest of Norway women in labour could get help from a birth-assistant, or a midwife.[225]

Dagny contacted a midwife when she was at the market in Bossekop, in December. The midwife had offered to come to Haldde and be there for a fortnight, but Dagny does not think this is a good solution. One of the problems about giving birth on the mountain is that, "There is no milk there, and it is a long way to the doctor, if it becomes necessary to have one."[226] No matter what she decides to do it will involve "The same amount of hard work, whether I take all the things I will need, or if I stay

---

[223] Letter from Dagny to her mother-in-law, 31.10.1912.
[224] Letter from Dagny to her mother-in-law, 23.01.1913.
[225] Aas 2011:10.
[226] Letter from Dagny to her mother-in-law, 15.12.1912.

here quietly and get them to send milk up every day." The midwife explains to her that it will be difficult for her to simultaneously hire a room and a maid to assist her when her time comes. Dagny tells her mother-in-law that her own mother thinks she should go down to Kristiania to give birth, but Dagny thinks that this would be too hard, both for her mother and for herself.

When it was good weather Dagny took the opportunity to go skiing. Archive: National Library of Norway - Photo collection.

Quite naturally, now that it is near to the actual birth, it is not equipment that she writes about in her letters, but she discusses where she ought to give birth. Dagny always keeps her mother-in-law updated about the situation:

> I was given information about what I should do in the spring. Either go to Kåfjord or to Bossekop. In Bossekop I could perhaps stay with, "Madam", that is the midwife, and I think I would prefer to do that, because now I see that although I am feeling well, I unfortunately have to take things more quietly than in the beginning.[227]

A beautiful white mountainous plateau for as far as the eye can see from the top of Haldde. Archive: National Library of Norway - Photo collection.

Therefore Dagny finally decides to go down to, "Madam", for the birth. Dagny is very happy when the midwife also offers her a room to rent in her house.

> I have had a letter from the midwife in Bossekop, and I can rent a room from her and a bed, but must have my own bedclothes with me, and for the person who will look after me. That will be quite a bit to take down. She says, I can perhaps get a midwife from Hammerfest to look after me, but I like the one in Bossekop. I know that she is clever and clean so I will put myself into her hands.[228]

Dagny needs someone to look after her after the birth, and help with the baby. She would have preferred it if her sister Arnhild had been there. A first birth without anyone she knows round her is challenging, even with

[227] Letter from Dagny to her mother-in-law, 1.10.1912.
[228] Letter from Dagny to her mother-in-law, 23.01.1913.

a clever midwife. That Arnhild decides not to come is difficult for Dagny to accept.

> Arnhild has now said no, and I do not know what her plans are, but think that our aunt has asked her to stay, because I believe that she is very kind and clever. I told her that I had not got further than arranging for a room and a bed at the midwife's house, but that I would be able to manage.[229]

Planning must continue with or without Arnhild. The next point on the list is when she ought to travel down to Bossekop. They have to evaluate what is best for her and the baby, and when it is practically possible to leave. She does not know when the baby will arrive, and is afraid of being alone for a long time. "Ole thinks I have to leave at the beginning of April, but I am not sure, just think if I go a month over my time and am away from here for so long? You see, one believes that one is indispensable!"[230] This is perhaps meant rather ironically, but Dagny knows that the work she does on Haldde is important for the small colony. Finally she makes up her mind about what to do, "I have now decided to leave here in April, which day it will be is impossible to say, but I have to send the clothes down during the week after Easter, and I need to find out when the drivers can do this. I must also have tolerably good weather for the journey."[231]

As always on Haldde it is the weather, not people, that decides when one is able to travel. A letter sent later shows that she left a couple of days before 31st March. As soon as she arrives in Bossekop she writes to her mother-in-law, and tells her about the journey down the mountain.

> Yes, now we are in Bossekop, and the journey went well, have described this in more detail in my letter to my mother. It is really very nice and cosy here. The walls in my room are light blue so that everything is stylish, and all the things you sent arrived safely, and are so lovely as well as being plentiful, enough for two says Miss Øverby. At the side of the bedroom I have a living room, which is extremely nice, and it is where we have our meals. Ole is sleeping on the camp bed you sent, there was no room at the hotel, at present, and I was happy that we were not parted.[232]

---

[229] Letter from Dagny to her mother-in-law, 23.01.1913.
[230] Letter from Dagny to her mother-in-law, 23.02.1913.
[231] Letter from Dagny to her mother-in-law, 13.03.1913.
[232] Letter from Dagny to her mother-in-law, 31.03.1913.

The Northern Lights dance over the observatory. Post card made by Krogness. Photo: Ole Andreas Krogness. Archive: National Library of Norway - Photo collection.

The date for the birth is at the beginning of May, so that Dagny knows that she will probably have to wait a month on her own, except for part of some weekends when Ole visits her. Every time he visits her, he has to come down from the top of the mountain to Kåfjord first of all, and after that travel 15 kilometres to Bossekop, we presume by horse and cart or boat. That he has installed a telephone for Dagny means that they are able to talk to each other, while he is on Haldde. Ole travels back and forth, and sleeps on the camp bed. This arrangement suits Dagny very well.

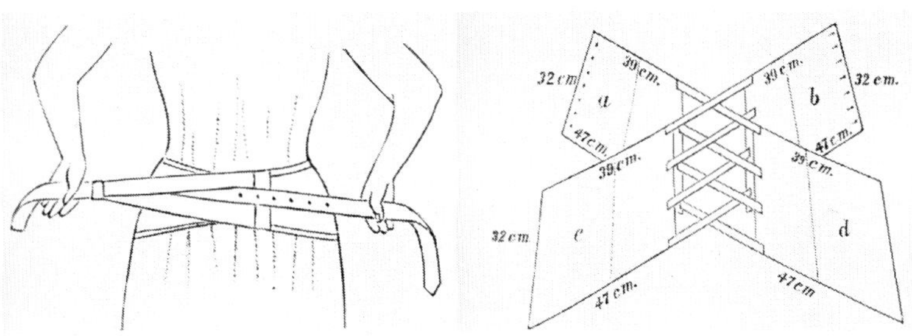


Doctor Greve's advice to pregnant women at the beginning of the 20$^{th}$ century. See Greve 1906:6

While Dagny is waiting for the birth, she has plenty of time to read and catch up on what Doctor Greve has to say about birth, and the end of pregnancy. He stresses the importance of hygiene. She has to make sure that those who look after her wash their hands thoroughly, this point is repeated a number of times in the book. The person who is pregnant should check and insist that it is done, and not blindly believe what the helper says. In addition to this the doctor and midwife should rinse their hands in "carbolic acid water or phenol water", to hinder the danger of infection.[233] The Norwegian Government was concerned about good hygiene and good nourishment, and gave out information about this, in an effort to prevent the death of mothers and their babies.[234]

Doctor Greve came with a lot of advice, also about clothing, in his book. He warns about garters and tight corsets, which were normal to wear then, as they prevented blood circulation. At the same time he understood that perhaps something should be done to compensate for this, and comes with other advice:

> An abdomen that hangs down should be held up with a stomach band [...]. A very good one is (...) Mrs. Fleicher Griebels' (Berlin). When the bands are tightened, they must sit under the tail end that holds it down.[235]

There is not much more that we can find from the letters about Dagny's "happy circumstances", before Ole radiantly writes to his mother on 7th May about the happy event, and that all is well with both mother and child.

> I am now sitting in "Aurora", our house down here, and have just come from visiting Dagny and little Ole Andreas (?). We have only just begun to speculate a bit about what we should call him, what do you think? A charming little fellow, who cries and drinks milk non-stop. I was somewhat surprised when Miss Øverby rang this morning to tell me that everything went well, and was fine, and that it was a boy that had arrived.[236]

Ole talked to Dagny the evening before the birth, and she had gone for a long walk and could feel no indication that something was about to happen. Labour pains started at 2a.m. and at ten past eight in the morning the baby was born, which was quite quick for a first birth. Dagny was in extremely good form, and asks Ole to take her a pair of boots when he visits her:

> I got down here a little before 7p.m., as I had to put some things in order before I left. She was still looking fit, but of course very listless and has had some pain since we talked on the telephone. The baby was to have milk that evening, and immediately started drinking very happily. Things went extremely easy for D, thank goodness, and everything is proceeding,

---

233 Greve 1906: 15.

234 Norwegian National Association for Public Health.

235 Greve 1906:6-7.

236 Letter from Ole to his mother, 07.05.1913.

as it should, so far. He has a lot of black hair and blue eyes and looks a lot like Guldbergenes, I thought straight away, but that is not to be relied on too much. Of course, hoist the flag up to the top of the mast.

Dagny Krogness on a visit to friends in Bossekop. Archive: Private, Ole Holbæk-Hansen

Now more than ever Dagny longs for someone from her home to support her during her post-natal period. Unfortunately, Arnhild is working for their aunt in Kristiania, for good wages and has not thought about travelling north. Nobody else comes either. The maid, Aminda, is still on Haldde and helps Ole and the assistants. Not long after the birth Dagny is told that the maid does not want to continue working there, even though Dagny has been very satisfied with her.

Dagny begins to get worried, and thinks the baby is restless, and that his navel is not healing, as it should. Dagny gets the doctor to check that things are all right, and he tells her to take good care of him, but the baby boy continues to be restless, screaming and crying as if he has a pain in his stomach. Dagny also thinks that his breathing is not as it should be, even though she takes very good care of him, but she is afraid and unsure. Ole is not worried in the same way, and thinks the boy is wonderful. Dagny writes to her mother-in-law about the situation, rather more directly than usual:

> You have perhaps heard that the boy has not been well, but now luckily he is better. His navel looks better and is beginning to heal. The bronchitis is at times better and then worse, and he brought up quite a bit of phlegm yesterday, which is not so good. His stomach is not quite right but he is much quieter now, so I hope things will calm down [...]. I think that when the boy is named it should be the name we have said, as we will most probably find a nickname for him later, if he is allowed to live.[237]

The baby is very quickly given the nickname Oa. Later in the letter Dagny writes that she is glad that the christening gown has been sent to them so early, in case something happens. The recipients now understand how everything has become too much for her, and now she is afraid that the baby may die. The family in Kristiania reacts quickly to this, and Arnhild is to travel up to her, and be there from the summer until the autumn.

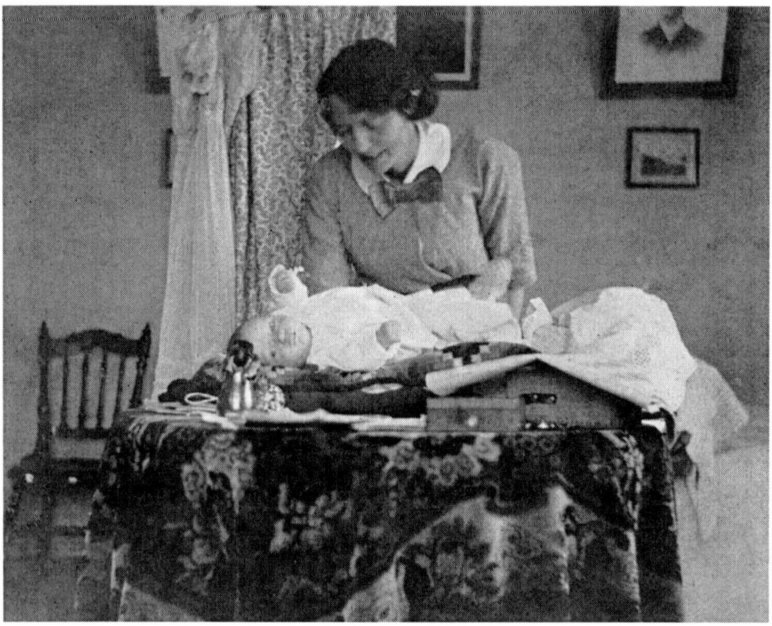

After the first hard weeks, Dagny calms down and enjoys being with her son. Archive: National Library of Norway - Photo collection.

[237] Letter from Dagny to her mother-in-law, 27.05.1913.

Ole is clearly perplexed about things that have happened, which he has not been completely aware of. He writes to his mother about how it must have upset Dagny's mother and sister, as they clearly believed that the baby boy was going to die. Ole is somewhat concerned that it became so dramatic:

Dagny plays with Oa, and takes especial care when they are on the roof of the shed. Archive: The National Library, Photo collection.

> [...] I have the impression that Dagny has frightened them quite a bit. First a telegram came with a categorical demand from the whole family, "Christen the boy" they stressed, and now I hear that Arnhild is coming here post-haste. At any rate, I was surprised about the first thing, the christening. Dagny thought it best to do what they said, so he was christened at home last Tuesday. I was down there on Monday, but I had not quite realized the problem. He was, as you have heard, christened Ole Andreas. That Arnhild came was really very nice [...][238]

Dagny realizes later how worried they must have been about what she had written, and she explains this to her mother-in-law:

[238] Letter from Dagny to her mother-in-law, 11.06.1913.

> I apparently frightened you, I believe, you know what such small babies are like. His navel is almost better, but the poor little thing will have to suffer with his bronchitis for a while. We rub his breast and back with oil and have begun to use woollen underwear […]. He looked very ill last week, but now we all think he has put on weight, […] and has put on a half mark [8 ounces or 250 grams]. […] I received a telegram saying that Arnhild was coming, and I am really looking forward to that, as to tell you the truth, things are rather difficult for me.[239]

After the first year, Dagny realises that it is important to be nearer to the family whenever she can. That is why Dagny decides to travel down to them in the summer of 1914. It will cost money and take time, but she decides to make this a priority. She writes to her mother-in-law, "We will see each other in the summer. Maybe you think the idea of travelling so far is a bit far-fetched, but I think it will do us good, so that we have time to thaw out ready for the coming years here in the north".[240]

Ole is alone on Haldde. Still full winter, 17th May 1913. Ole Andreas Krogness. Archive: National Library of Norway - Photo collection.

[239] Letter from Dagny to her mother-in-law, 10.06.1913.
[240] Letter from Dagny to her mother-in-law, 27.12.1913.

# CHAPTER TEN

# PRACTICAL HARDSHIP AND ACADEMIC CONTROVERSIES

Størmer and an assistant take photographs of the Northern Lights from Bossekop. Haldde is in the background. Photo: Carl Størmer. Archive: Norwegian Museum of Science & Technology

*It is with mixed feelings that I go to Haldde again. I am mainly worried about our little boy, but one only hopes that everything will go as it should, and be better than we think and can hope for.*[241]

When the baby is born Dagny is quite naturally mainly concerned about him in her letters, but now and again she is able to squeeze in a bit of information about Ole's work. "Ole is writing an annual report for the

[241] Letter from Dagny to her mother-in-law, August 1913.

Government now, and is working with a lot of financial accounts, so you will not get any letter from him at the moment."[242]

Ole must have been interrupted while working with this, because according to Devik, the Governmental Department had commented to him during an application process, that the annual statements from Haldde for 1912 and 1923 had not been delivered on time. The bureaucrat concerned had said, "I will tell you, young man, that the most important information for a scientific institution is the annual report!"[243]

Ole takes note of this and that he does not have enough time for his scientific work. Jobs such as clearing snow, chopping wood, keeping the machines working and emptying the toilet takes a lot of his work capacity and time. He includes this information in his annual report to the Government, and tells his mother about this:

> I have suggested that we should also appoint a general workman as soon as possible as "caretaker"; which is an excellent suggestion, but probably nothing will come of it, because they have to take the budget into consideration. There is quite a lot that is needed and now we have room for one or possibly two helpers.[244]

The letter shows that they have been given permission to enlarge the house on Haldde, which will improve their existence considerably. It is especially a relief for Ole, because a larger house can strengthen his application for more help with the work. Dagny comments in her letter, which is sent about the same time, "At last Ole has written a letter. Yes he has a thousand things in his little head which is, by the way, covered by a beard and lots of hair."[245]

---

[242] Letter from Dagny to her mother-in-law, 06.09.1913.
[243] Devik 1971:75.
[244] Letter from Ole to his mother, 14.09.1913.
[245] Letter from Dagny to her mother-in-law, 13.09.1913.

Back on Haldde with the baby in the cradle basket. Archive: National Library of Norway - Photo collection.

The newly fledged father on Haldde, 18th May 1913. Archive: National Library of Norway - Photo collection.

It became even harder to live on the top of the mountain with a small baby during the autumn of 1913. There was all the extra washing, and the bad weather also meant it was difficult to dry anything. When the maid left it was difficult to get another one. It also took longer than they had thought to enlarge the house. Dagny reports that Ole is often worn out and tired. She herself complains about the loneliness and absence of someone to talk to. She longs to move south. When they get a visit from Olaf Devik's younger brother Karl for a week, it is quite an exhilarating experience for them both. Karl had come as a representative for Birkeland, and went around taking lots of photographs and making rough sketches, Ole writes and tells his mother:

> It was an extremely welcome visit especially as I had so much on my mind that I should have written about, but because of the quantity of it I never got round to actually writing. When Karl was here he got all of it reported orally and wrote down things he should remember, and just that covered a lot of paper. He had to promise to take all his sketches and photos and go up to you in Holberg Street, and give you a short lecture about everything in detail, so that you can hopefully see what the new things look like up here.[246]

In the letter, Ole also tells his mother that Olaf, who had been his best man when he and Dagny were married, has also got married. The bride's name is, Dagny Hansen, and like his own Dagny, she had been an employee of, "*Studenterhjemmet*" (Students' lodging-house) in Kristiania, for a while. They have sent a telegram from Haldde to the bridal pair with the following text; "The greatest happiness that can be wished, we wish you, Dagny, Ole and little man." In the same letter to his mother Ole also comments about his assistants:

> The new Larsen has been in charge on his own up here while we were away and things went quite well. I have, by the way, a feeling that he will be leaving quite soon. Birkeland had halfway promised, not to say, completely promised, that Parliament would raise his wage from 1200 to 1800 and they have not done so. He is rather a strange fellow, but does a good job, and that is the main thing. I think that Hoel has shown great improvement; perhaps the higher wage has cheered him up.

Between December 1913 and August 1914 there are no letters. What happened to the small family in that period is not clear. It looks as if Dagny left Haldde with the baby and lived in Kristiania. In July 1914 war

[246] Letter from Ole to his mother, 14.09.1913.

broke out in Europe. It is not unlikely that this may have been why Dagny wants to spend some time with her mother. Although the most probable reason for leaving Haldde earlier than planned is the slow building process with their house on the mountain. Ole wanted to see the eclipse of the sun and planned to go to Kristiania in August. However, Dagny had just returned to Haldde and there had not been much letter-writing for a while, but on the 22nd August, Ole makes a big effort, and writes to his mother:

> As you can see I am still up here, which means that there will be no eclipse of the sun for me. It was difficult to get away, as well as there being a lack of money, as I have not received anything from the Government. They seem to be so confused down there, that things get muddled up for them.[247]

He does not seem especially concerned about the war. In his usual somewhat humorous way he comments about the situation:

> There is a tremendous commotion down there. Up here there are a number of strange rumours. One day it is the Russians in our waters, and then the newspaper tell us that it was in the Dardanelles. Geography knowledge here is somewhat poor, so that they have perhaps thought that the mystical place mentioned, was somewhere in Norway.

It seems that for Ole the war has come as rather a surprise. He had read in the newspaper that the Prime Minister Gunnar Knudsen's attitude, was rather aloof, just before the war broke out. Under the title "What is Happening?", the newspaper referred to the Throne Debate where Knudsen had stated that the powers that be, were not interested in the downfall of Norway, but that they were more interested in Norway's independence. The situation was not exactly that Norway would sacrifice everything on a military defence. Later there was an announcement saying that the Government Cabinet had decided that, "As far as Norway is concerned we will remain completely neutral."[248]

In a letter to Birkeland in November 1914, Ole mentions his attitude toward Størmer. He writes in English because of the war to avoid the letter being stopped and censured. Ole hopes that Birkeland will excuse his, "surely terrible english". It is an urgent matter, which is why he writes before he has Birkeland's address in Egypt. His reason for writing is that both Vegard and Størmer have individually asked to borrow a series of magnetograms from the date that they had done their observations of the

[247] Letter from Ole to his mother, 22.08.1914.
[248] Hølaas 1945, about 1914.

Northern Lights. Each of them wants ten magnetograms.[249] Ole refers to what Birkeland said the previous summer, that he was in agreement that these two could have copies. This has made it difficult for Ole not to hand them over, although he is not completely sure if he can legally do so. He thinks that perhaps he can give them a couple of the copies, they have asked for, and wait to hear what Birkeland has to say. He has written to Olaf Devik and told him what Birkeland said the previous summer and asked for advice about this. Ole continues in English:

> From copies received, Störmer has on the 11–12 mars 1913 according to a referat in the journal – and through a letter from him, just received with the request for the magnetograms – arrived at the conclusion that the sign of the corpuscles was positive. The grounds for this conclusion he has not mentioned.
>
> This day is the very same as that on which I last year as you will remember, arrived at the same conclusion. Under these circumstances it seems to me that I can hardly wait longer with the publication of this observation.[250]

It appears that Ole has come to a similar conclusion as Størmer, and that it should be published as soon as possible. The problem is that it will also confirm Størmer's theory, and possibly conflict with Birkeland's own theory. There are two ways of formulating this in a publication, Ole says. In an article they can either admit that the observed light movements may be caused by the movement of the actual rays, and that it is likely that these correspond to the positive rays. Or they can claim that if the light movements were caused by the relocation of an entire ray system, such a conclusion cannot be reached. Moreover, they can point out that one observation cannot be considered as the final result, but perhaps later observations carried out by others, can give interesting results. Ole clearly wants to avoid reporting about a discovery that agreed with what Størmer had found. He did not want to support Størmer's critic of Birkeland's theory, and become involved in a conflict between the two of them.

[249] Magnetograms are graphical representations of measurement results.
[250] Letter from Ole to Birkeland, November 1914.

An exhausted Ole, absorbed in his work. Archive National Library of Norway - Photo collection.

In a detailed written account, Ole tells about additional interesting observations he has made of wind and temperatures at varying heights, what the theoretical consequences are, and what conclusions one can draw from this. Ole has sent a draft of this article, about the observations of the Northern Lights to Devik, and if he agrees with him about this, he will send a copy to Birkeland, as soon as possible. Devik's sympathy in the controversy between Størmer and Birkeland is evidently on the same side as Ole.

It is well known that both Vegard and Størmer agreed with the British critic of Birkeland's theory about the Northern Lights. It was this that had been the basis of the earlier mentioned newspaper debate in which Størmer and Birkeland did not agree. Vegard argued that the positive ions are a dominating contribution to geophysical phenomena. That is not in accordance with Birkeland's theory about the Northern Lights. Despite their disagreement Birkeland continued to support Vegard's study of the Northern Lights, but they never published anything together.[251]

Ole's letter touches upon the internal research situation, obligations, agreements, cooperation and competition between researchers working in the same research field, and which of them publishes the most interesting results from their projects, and which results are most credible. It seems

[251] Egeland 2012:161.

that they are well aware of what each researcher in the geophysical environment has discovered, and that this information contributes to furthering scientific knowledge about what is taking place in the stratum of air; independent of researchers' intern controversies. The Haldde researchers are central, active participants in the geophysical research milieu in the Scandinavian region. There are tight bonds between the summit of Haldde Mountain in the North of Norway, and the central scientific research milieu.

It is a long letter mainly about the research. The most interesting discovery is that in the winter there is a, "monsoon system", in this Nordic region, a special wind near to the earth's surface and a north-westerly wind in the higher stratum that come from the, Golf Stream, and blows towards the continent. This wind is so low on the summit of Haldde that most of it can be found within the same wind current.[252] Ole regards this as extremely interesting.

At long last the Krogness family had got some company on the top of the mountain. The Lukkassen family had moved up there, with a dog. Archive: World Heritage Rock Art Centre - Alta Museum

[252] Letter to the Royal Church and Education Department, 08.10-1914.

He is working on a dissertation about whether it is possible to establish any connection between the magnetic perturbations[253] that he has registered, and wind conditions in the region. This is relevant to the planning of a storm warning system in the Nordic region. So far there has only been storm warnings for areas south of Bodø. Unfortunately, Ole has not had the opportunity to study how the wind conditions in the Nordic region fit into the total meteorological situation. He has been too busy.

Finally, in his long letter to Birkeland he mentions a bit about daily life. Lukkassen, from Kåfjord, has been given the position of assistant, and "handyman", and has moved up to the top of the mountain with his wife, three daughters and a cow. He also mentions that Hoel is about to enter the holy state of matrimony that winter. Things will then be brighter and livelier on the lonely summit. The house has now been enlarged and there is plenty of room for everyone.

The main thing now is to get an answer from Birkeland about what he thinks about Størmer and Vegard's question. Ole asks that the answer should be given via Devik, who can probably telegraph it to Ole.

Oa had playmates when the Lukkassen family moved into the house. Photo: Ole Andreas Krogness. Archive: National Library of Norway - Photo collection.

[253] Term describing a disturbance in a single and well-defined state of motion. Store Norske leksikon (2005–2007). See https://snl.no/perturbasjon_-_fysikk

This takes place during the period when the mathematics Professor, Carl Størmer (1874-1957) was living in Alta, taking photographs of the Northern Lights. The densely populated area of Bossekop had become a Mecca for studying the Northern Lights in Europe. Størmer had earlier developed a mathematical model to go more deeply into, and to formulate Birkeland's work, which had fascinated him for quite a while. He had studied the height of the Northern Lights from Bossekop with the help of photographic technics. He could now with assurance claim that the lower edge of the Northern Lights in most cases was about 100 kilometres above the ground, and sometimes the top edge was several hundred kilometres up in the atmosphere.[254]

It was Størmer together with Ole, who had in 1909 developed the world's first usable Northern Lights' camera.[255] Probably it was this camera he used in Bossekop. Together with the earlier mentioned "fusspot", Lars Vegard (1880-1963), he had made observations and measurements of the dancing light phenomenon in the Arctic skies. From Størmer's two months' long stay in Bossekop, more than 2500 Northern Lights photographs were carefully analysed. The result of this work was the first reliable height statistics of the Northern Lights.[256]

Vegard was mostly known for having identified and mapped out the colours in the Northern Lights.[257] He had begun his career as an assistant to Birkeland in 1905-1907. The work he did later was regarded as being pioneering work. After just a few years' work he had accurately identified and mapped out that there were about 40 different colours to be seen in the Northern Lights. Vegard continuously made new discoveries during the years from 1910 and onwards.[258]

The mountain summit above the small village of Kåfjord in Alta fjord was involved in research activities and controversies, with the elite Norwegian scientific milieu of that time. Following the communication between Ole and the other researchers, gives us insight into what was happening in the Norwegian scientific milieu, which consisted of prominent personalities. Each of them in their own individual way has contributed to laying down the foundation for today's research at universities and research institutions.

A month after Ole's comprehensive letter to Birkeland, Olaf Devik has visited Haldde, maybe because of the need to discuss the circumstances

[254] Friedman 2004:131.

[255] Egeland 1012:29.

[256] Egeland 2012:31.

[257] Brekke & Egeland 1979:74.

[258] Brekke & Egeland 1979:75.

surrounding research conditions that Ole had written about in his letter. We do not know what feedback they got from Birkeland, or what they discussed together about what they should do, in connection with the questions from Størmer and Vegard. Ole writes about this to his mother:

> Devik has now gone up to bed, and I am making use of your method with the quiet evening hours for writing letters. It has been a great experience having him here for these days. His wishes have now been fulfilled, namely both to see the most magnificent weather that we can offer and a couple of tremendous hurricanes. He will be able to tell about this when he gets back to the capital together with this letter. He has promised to call you before he comes on a visit, so perhaps you could send Arnhild with a message to Mother Guldberg. She is also most likely interested in picking his brain a bit.[259]

It was through letters and visits that contacts were kept in a work and family network, in which women were active participants. Dagny's mother would receive an invitation from Ole's mother to visit her when Devik was back to Kristiania. Arnhild, Dagny's sister, would also be able to appease her inquisitiveness as to how the family was coping with living in probably the world's most harsh weather-exposed locality.

---

[259] Letter from Ole to his mother, 17.12.1914.

# CHAPTER ELEVEN

# THE MAIN HOUSE EXPANDS

The extensions to the house gave work for the men living in the village. Photo: Ole Andreas Krogness Archive: National Library. Photo collection.

*We take care that there is no carbon monoxide, we are not sure if it is the coke that is bad, but we have now been so frightened by what happened to the assistants.*[260]

The episode with the carbon monoxide poisoning in the autumn of 1912, had really frightened them all. The letters tell us that under certain wind conditions they also had problems in the main house, when the fire was alight all night. One of the explanations was that the coke they used was of bad quality. Dagny writes that she noticed this when the new supply of coke arrived, as it was much better. When it turned out that the

[260] Letter from Dagny to her mother-in-law, 10.11.1912.

observatory was built incorrectly, and that there were no ventilators in the rooms, this meant that the assistants could not have a fire overnight, and therefore could not live there. On the other-hand, in the main house there were ventilators, which they could have opened except when the storms were really bad. Even so, when the weather allowed it, and it was still cold, they could have a fire burning overnight in the main house. “I have tried to have a fire all day and all night, as it is so cold here during the night, and when I move my head on the pillow I feel it turns to ice”, Dagny wrote to her mother-in-law a month after they had arrived on Haldde.[261] One can just imagine how cold it must have been without any heat from an oven, up in the “Castle”, which was what they called the observatory.

When the assistants could no longer live in the observatory, the main house was overcrowded, and it meant that they lived even closer to each other. This situation was later used as a reason for enlarging the living area. Dagny tells her mother-in-law about the challenge of having guests for Easter in 1913.

> Four guests are coming here for Easter and I have been thinking about where we should put them; of course we have the assistants’ empty rooms. Hoel has moved down to our house, as the carbon monoxide is so impossible up there. I hope we get some money to build, because now it is impossible to keep any order in the rooms, when they are used for all kinds of things.[262]

Ole had applied for money from The Ministry of Church and Education giving the following grounds for the application. “Earlier, and quite a number of times previously, I have informed you that we require more room to accommodate the staff, especially as the old observatory has now shown itself to be completely impossible to use in the winter.”[263] Building started in the spring of 1913 when they had all moved down to spend the summer months in Bossekop.

---

[261] Letter from Dagny to her mother-in-law, 10.11.1912.

[262] Letter from Dagny to her mother-in-law, 17.03.1913.

[263] Letter from Krogness to the Royal Church and Education Department (KUD) 20.09.1915.

The house on Haldde looked like this, before work started on it. Photo: Ole Andreas Krogness. Archive National Library, Photo collection.

Autumn 1913 the building work was still in process, even after the couple, baby, assistants and maid had moved back up there. Dagny reported this situation to her mother-in-law.

> Yes, here it is full of workmen, the whole Castle is full as well as the attic, yesterday they made a hole through the roof into our house, in the dining-room there will be a door into the large work room, but that has not arrived yet. Yes you can imagine that I long for everything to be neat and tidy but I think it will continue to be like this until Christmas this year.[264]

The work on the building did not go as quickly as they would have liked, and the house stood more-or-less unfinished in 1913-1914. The builders only managed to complete four rooms, which enabled them to be used that winter. There were two reasons for the delay, one was due to unfortunate circumstances for the builders, as their leader had to go back to his original job because of illness, and secondly because the winter started early and therefore the workers had to stop work in the autumn of 1913. The house was finally finished in the autumn of 1914. Its final size was now about 19 square meters, with a cellar and two floors.

[264] Letter from Dagny, to her mother-in-law, 06.09.1913.

The new building was built of a lighter stone than the original one. The stones used, were from the scree down below the house. Photo: Ole Andreas Krogness. Archive: National Library, Photo collection.

Originally the ground floor in the new building was for the Director and his family, but when later, it was decided that Devik should move up there, further changes were made. The office was divided into two parts, a living room for Devik and the other smaller part an office that could be shared. Additionally, the Krogness' guestroom became Devik's bedroom and the "maid-servant's room", became Devik's kitchen. The Krogness family's flat was now three rooms and a kitchen. On the second floor the Lukkassen family and Hoel each had their own flat, consisting of two rooms and a kitchen, and there was also a room for the maid on that floor. The Lukkassen family's flat was above the Krogness' flat and Hoel's was above Devik's.

In the course of the first years a device was constructed in the machine house, which was extremely labour saving for when they had to melt snow. A large funnel stretched down into the machine house, and the actual mouth of the funnel was beside a large snowdrift, which made it easy to shovel snow into it so that it went down into the pan below. This meant that they did not have to carry small quantities of snow into the kitchen to melt in copper pans. To be able to have water throughout the spring they had built a cement cistern, but if this dried up they had to fetch water from a nearby lake, which was hard work.[265] It was due to a lack of water that the family moved down from the mountain in the summer.

The floor being laid in the new part of the building. Archive: The World Heritage Rock-Art Centre, Alta Museum

In addition to the observatory and the main house, other small buildings were built around them. In the machine house they had placed the laundry room and what they called a little "*kvenbad*", a type of sauna, which was a Finnish custom brought to the area by Finnish immigrants. In one outhouse there was a larder, where most of the food for the four families was kept. Eventually they built a room in which to dry clothes and other things, which had been something Dagny had really missed during the first years on Haldde. There were also two small rooms in the house, one for the toilet and another for the firewood.[266] A small barn for the cow with a section for the hay, needed during the winter, was also built.

[265] Krogness to NVE about Kåfjord's hydro-electric station, 24.09.1925. In Alta Mueum (1999) Use and Restoration plan for, "*Nordlys*", Obsevatory.
[266] Krogness to NVE about Kåfjord's electricity works, 24.09.1925. In Alta Mueum (1999) Restoration plan for Nordlys Obsevatory.

One of the assistants looking for the Northern Lights on a cloudy evening. Photo: Ole Andreas Krogness. Archive: National Library, Photo-collection.

The tunnel had been extended into a longer underground passageway that was connected to all the buildings, so that they could move between them when there were storms outside. To make it easier for transport to get up the last steepest part of the mountain to the house, a cable was extended down to a plateau 600 metres below. This was driven by electricity from a petrol-driven generator or an accumulator battery. The actual roadway up from Kåfjord was about 8.5 kilometres.

The Northern Lights' Observatory also had a couple of properties down by the fjord. They owned a house in Kåfjord, which was 96 square metres plus a small piece of land attached to it. It was here that the families could live when the children started school, and Lukkassen, who was the caretaker and assistant, made use of this. They also had a house in Bossekop with a ¼ of an acre of land.

The reason for enlarging the building was first and foremost due to the problems they had experienced with the carbon monoxide poisoning, but also because of increasing activity. A lot was improved in the new building, but despite this, the problem with the carbon monoxide persisted. The wind and smoke drove downwards and into the building during certain weather conditions. It is true that after a while they became more experienced about when they had to be careful, but the problem remained.

The house is finished and stands proudly in its splendour. Archive: The World Heritage Rock-Art Centre, Alta Museum

Ole, who was not just a researcher, but also a capable inventor, tried every kind of solution in an attempt to solve the carbon monoxide problem. He made a fan, which was placed in the stovepipe, and let the wind drive it to get the carbon monoxide out. It functioned except that two of them broke, but as he said that was due to a "production error". Another invention came to light after Devik had moved in. This is how the now elderly Mrs Devik explained it:

> During a carbon monoxide storm, the small kitchen was filled with soot extremely quickly. It settled everywhere and we could pick soot off the walls […] Then father [her husband] got one of his good ideas. We had kept the oven burning constantly day and night, and it resulted in a hole in the pipe and so he fitted a pipe into this, like an arm sticking out, it was about so long, right over the end of this pipe was to be a paper bag. We did not get much further until we had a larger paper bag. Nothing came into the room. It was effective when the wind blew in carbon monoxide. The two Krogness children and the others from upstairs came into us to save themselves. When a really hard gust of wind came you can just imagine how it crackled in the paper bag, but nothing came into the room. Ingenious![267]

This was not of help in the long run. The carbon monoxide was, and continued to be a problem. It could also be very cold. Dagny and Ole's daughter Synnøve Holbæk-Hanssen, described it like this, in an article for the newspaper, "*Finnmark Dagblad*", 19th January 1981.

> During the coldest part of the winter we went to bed wearing mittens and woollen hats. The stonewalls were thick, but even so the snow could get through. The adults kept a fire in the oven, but there could be high pressure when the wind blew strongly, which forced the carbon monoxide downwards. It was better when we got electricity from the closed down mine in Kåfjord.

Synnøve, who had asthma, and was often ill during her childhood on Haldde, illustrates here the dilemma of needing a fire to keep warm, but which was difficult to have during storms and other bad weather, because if there were extra strong winds, there was the added risk of carbon monoxide being pressed downwards into the house.

---

[267] Conversation with Grandmother 27.08.1985. Dagny Devik and her children Ingeborg, Dagmar, and Ole.

The carbon monoxide problem was an important issue and a main reason why Ole began to work with the idea of getting electricity to Haldde. Dagny informs her mother-in-law about this, "Ole is struggling to get electric power for us next year, but this is not as easy as it sounds."[268] A lot of time was used on writing letters, applications and negotiations before this work had a happy ending.

This is what it looked like on Haldde after rebuilding in 1913. Archive: Haldde Observatory's Archive in the National Archives in Tromsø.

[268] Letter from Dagny to her mother-in-law, 12.02.1916.

# CHAPTER TWELVE

# NEW PLANS

The work situation is easier for Ole with a new collaborative partner. Archive: National Library of Norway - Photo collection.

*We are lucky that everything is fine here, when so many are, at present, having such a hard time.*[269]

Dagny was conscientious in writing letters to her mother-in-law, even now when she is kept busy as the mother of a baby, and Christmas was an especially busy time.

> Dear Mother, Many thanks for your last letter. I have a somewhat bad conscience because I have not thanked you for each thing you sent, but there have been so many letters to write, and I had to write all of them. I

[269] Letter from Dagny to her mother-in-law, 07.01.1915.

> am trying to teach Oa what animals say in the little book you sent, it is excellent that there are material pieces too, but I hide them from him, as he is quite awful about destroying things.[270]

In the same letter she continues, "The cloths you sent were fine, the one is good for polishing silverware now and again." On that mountain summit a high living standard was kept, regardless of whatever else happened.

While the war makes things unsafe in Europe, there are developments and enthusiasm in the Arctic research station. From being a deserted and lonely place for a young couple, a baby and a maid to live in 1912, plus some rather impossible assistants, Haldde is now almost a small mini society. Also new aspects in the research emerge. Creativity blooms and activities expand. Dagny, who for quite a long time has longed to move south, writes to her mother-in-law:

> You have now heard Devik tell you about our plans. You know I would prefer to come to Kristiania, but I have to accept that Ole does not want me to go and take responsibility for the journey on my own. It was apparently Devik's plan to take over leadership from the autumn, but Ole does not want that to happen. He feels that he has not got enough results yet, so that it might take even more years. I think that you must come and visit us, at least once. And there is no good position for Ole down south at the moment either.[271]

That Olaf Devik is interested in taking over Ole's job on Haldde can also be read about in another letter. It seems that in May 1914 Ole had applied for a position in Horten. His close friend Devik had written about it in a letter 13th May 1914 to Professor Vilhelm Bjerknes (1862-1951), the well-known originator of modern weather forecasting.

> I would like to ask you straight away what you think about a case, which is now of interest to me, although at the moment it is a private matter and is therefore confidential.
> The situation is that Krogness is thinking about moving south from Haldde, mainly because he has his hands full up there and it does not look as if there is any possibility of him being able to work with the material he has collected, in the near future, but he would have a chance of doing this if he moved south. Personally he is extremely pleased with the work up there. It is of course of great importance for the observatory that the result of this work is published, and this in itself should be significant enough.

[270] Letter from Dagny to her mother-in-law, 04.01.1915.
[271] Letter from Dagny to her mother-in-law, 07.01.1915.

> Earlier I told Professor Birkeland, when he asked me, that I was willing to take over from Krogness if necessary, and I still mean that, as I personally want to convince myself about the observatory's usefulness and possibilities, and am not, by the way, afraid of the loneliness up there.[272]

In further correspondence with Bjerknes, Devik suggests that Krogness should try to get a position at the University in Kristiania. Devik feels that for Ole to take a job in Horten will be the wrong professional direction for him to take, argued from a scientific point of view, because it is outside, "the main scientific milieu". Even though Dagny has expressed her loneliness and unhappiness and that she longs to move south, neither she nor her son are mentioned in the letter, despite such an argument probably weighing heavily on Ole's conscience

"Besides, the most important personal factor in the question concerning Haldde, is Birkeland, and he is still out of reach," Devik writes. In all probability Birkeland is still without a permanent address in Egypt. That he plays a personal role in the dilemma about what they ought to do, can mean that it will cost Birkeland his reputation if Ole chooses to leave the observation post on Haldde. Apparently Ole and Olaf had got together and discussed this, and arrived at other and better solutions to the problem. Krogness should remain in his post, and continue with his research as before, but new plans are made.

The plan is to start registering the wind conditions by establishing measuring stations in several different places in the northern region, both on high mountains and by the sea. It is for this purpose that Devik is going to move to Haldde. This will undoubtedly be a positive contribution to the research work. It will increase its scientific reputation, create new cooperation possibilities and strengthen the position of those who have toiled through hard-working days and nights observing the Northern Lights. The Ministry has already been informed about the advantages of the planned research, how it will be carried out, and what the researchers will need of instruments and other resources. With regard to the social aspect, Ole hints that to be ordered to an isolated existence, can be in the long run quite a strain, even if the work itself is interesting enough. Therefore the plans that have been made will represent an improvement for a number of different areas, in their present circumstances.

Also their communication with the world around them would be improved. Ole writes to his mother, who he has communicated with via telephone, and explains the telephone connection, which is not that easy to operate.

---

[272] National Library, Handwritten collection, Oslo.

> The line between Bossekop, Trondheim and Kristiania is still not open, so that those of us who are on the line between Tromsø and Alta have a much poorer line on which to speak. It is the first time that one can speak from Finnmark at all, otherwise it is only from Tromsø that it is open, and even the Telegraph director himself has apparently not tried to use it [...]. Next time we could perhaps try, that when you want to hear, you press the button you use when you are telephoning to someone, but not when you are speaking. The fact is that then one can hear considerably better.[273]

That a telephone line was opened was a milestone. Nevertheless, telegrams were still used when letters would take too long. It was especially for important questions that telegrams were used between the capital, and the top of the mountain. Ole's mother was a constant central figure in the exchange of information. Ole continued in his letter:

> Just got a telegram from Devik saying that he was very excited about the plan. Now the big question is, whether I should take a trip down south and confer with Devik and Sandstrøm about all of this. I cannot deny that I would like to do that, and it would be favourable, so I am seriously beginning to think about it.

Photographs are taken outdoors and the view could not have been better. Photo: Ole Andreas Krogness. Archive: National Library of Norway - Photo collection.

[273] Letter from Ole to his mother 21.02.1915.

Johan Wilhelm Sandstrøm (1874-1947) was one of the eminent physicists to be found in the geophysical milieu in the Nordic region. He was a meteorologist and oceanographer, and was engaged in how heating and cooling affected the sea currents. For many years he was linked to the University of Bergen, acting as an assistant to Professor Bjerknes. The plan that Ole Krogness and Olaf Devik were occupied with, also dealt with the research area that lay within Sandstrøm's field of interest and competence.

1915 was an eventful year in many ways. There were significant events, both positive and negative throughout the whole year. Research continued, but Ole's efforts at taking photographs did not always turn out successfully. In the spring he writes to his mother, "Thank you for all your hard work with the film, it was an exceedingly unsuccessful affair, as it was impossible to distinguish anything. It was a little better with the stronger lighting, but even so…."[274]

In the same letter he tells her about a very happy event, "Am now delighted that, the Government agree with Devik's and my plans. Just hope that the Budget Committee also support this, so that it will be more lively up here next year."

The Parliamentary proposal number 68, "About establishing a grant for the preparation of a weather-forecast system for the northern part of the country," was presented to the Cabinet Minister 19th March 1915. A condition for this was, that Olaf Devik should be the leader for the preparatory work. Both Bjerknes and Birkeland sent their warmest recommendations. The Budget Committee approved the proposal, and Parliament passed it unanimously 27th May 1915.

Ole is extremely excited and writes to his mother, who has already been informed, and has rung to her son about the result:

> I am so pleased at the thought of having Devik up here and after what you told me on the telephone there is absolutely no doubt about it being passed. Now in connection with the question about Sandstrøm, Sweden and Russia, I wrote to Devik and asked him whether we should try and establish cooperation with Sweden and Russia, and now I see that the newspapers have found out about this.[275]

It is possible that the reason that Ole wanted cooperation with Russia was in connection with earlier mentioned observations and measuring of stratums of air over the Nordic regions. He had reasoned that the sea

---

[274] Letter from Ole to his mother 25.03.1915.
[275] Letter from Ole to his mother 04.05.1915.

outside the coastline was kept warm by the Gulf Stream, which in turn led to the formation of a powerful wind system. Heavy cold air from the continent moved in a northerly direction, and filtered out into the sea. To compensate for this continental wind there had to exist a countercurrent from the sea and towards the continent, and he called this a kind of monsoon wind. This had to be researched. Ole believed that before one could draw any absolute conclusions, long-term observations needed to be carried out. Thorough investigations of the meteorological situation covering a wider area also needed to be examined, as well as the conditions to be found in the higher stratum of air. There was a lot of work for the Haldde researchers to do.

Birkeland endorses the new plans on Haldde. From Egypt he writes to Devik:

*Dear Devik,*
*I am extremely pleased that already in July you will be moving up to Haldde. I am certain that Krogness and your work will be a great achievement for Norway. I would have liked to talk over a number of things with you, but I am unable to leave here this year. One piece of advice I will give you before you go, just in a general way. Use your own eyes, ability and reasoning, and do not let them in any way be influenced by what others want you to see. When you work conscientiously and diligently, you will also be satisfied with the field of research that you and Krogness have planned to work with. Glück auf!*
*With kind regards to you, and Krogness.*
*Your devoted Kr. Birkeland*[276]

A new letter to Birkeland from Ole in August 1915 informed him that he has taken him seriously about contacting Sandstrøm. It is a long letter. He calls it a, "little opus". He has so much on his mind after a journey south. An article has been published and it has received some reactions from the scientific milieu. Birkeland is informed about its contents. Vilhelm Bjerknes has commented on some negative results in a letter or telegram, which Birkeland had been rather taken aback by:

> I can also see that there has been some nonsense about Bjerknes's statement about "the negative results". I cannot say anything else than I too was somewhat surprised; I had written something quite different, but it is also understandable that he, as a, "dynamic meteorologist", one could say, found the greatest satisfaction in the "monsoon".

[276] Letter from Birkeland to Devik 30.05.1915, ref. in Devik 1975:110-111.

Bjerknes has quite clearly been positive to Ole's plans to research the existence of the monsoon wind. Bjerknes had also expressed something about all of it being put into Devik's hands. That was, "quite strange", Ole writes:

> […] as all it was about was us two collaborating. This was not unkindly meant, apparently, and it has not upset anything, as between Devik and me there are absolutely no doubts with regard to this, but it is possible it is a result of those I can gladly call intrigues, which most probably originated from the blessed A.S.S.[277] It is of course quite comical how this has held onto the old grudge about Haldde, "unter uns gesagt". Quite obviously every kind of means has been used to get at us.

It is not known what old grudge is being talked about. There has been criticism over a scarcity of results from Haldde, and this has been mentioned in the material we have, in a number of places. This was especially concerned with a deficiency of public information about what kind of discovery has been made. Ole felt that this was unjustified criticism, as they had published all the significant discoveries they had made.

A harsh and critical comment made by the meteorologist, Nils Johan Føyn, was written in an article in, "*Norsk Fiskeritidende*", the Norwegian journal of fisheries: "The meteorological work seems to be neglected, so far as one can see. Actually one knows very little about what has been accomplished, as no observations have been published."[278] In the same article, Føyn also criticizes Devik, for not having any experience with practical weather services, to allow him to carry out meteorological studies in the northern region. This criticism was because the work was to be carried out from Haldde, and not from a station in Bossekop, which Mohn had suggested:

---

[277] Unknown initials.

[278] Føyn 1915:148.

A small filmstrip from the researchers' filming of the Northern Lights, on Haldde. Archive: National Library of Norway - Photo collection.

Both Ole and Olaf had given well-founded answers to the criticism from Føyn, in their individual articles. Ole refers to these when addressing Birkeland, "I had, to begin with, used much stronger words than those that were printed, but because it was enough to refer to facts the more offensive expressions were rubbed out."[279]

During Ole's visit to Kristiania the two peaceable researchers had decided to take direct contact with Føyn, in his office at the Meteorological Institute. During the visit they were in agreement about not wasting time insulting each other, and instead try and cooperate as much as possible. Ole hopes that the result of the visit will have a positive effect on Føyn and that he comes to his senses. "Hopefully all is now well, with regard to that," Ole writes to Birkeland.

[279] Letter from Ole to Birkeland, August 1915 – Alta Museum collection.

The correspondence with Birkeland was important for both parties. Archive: National Library of Norway - Photo collection.

He continues the letter and tells about the visit to Stockholm where they met Sandstrøm, "an extremely splendid man!" They had been given a lecture on the significance of the mountains, in connection with the formation of storms, both in Northern Norway and during different geological era.

> He even, believe it or not, explained both the ice age and tropical climate, here in the north and everywhere else, based on the fact that when there were high mountains, the wind and climate results were subsequently as one might have expected, but when there were no high ranges of

> mountains, it was calm and the solar energy turned warm; quite extraordinarily gratifying.

Olaf Devik writes in his memoirs, later in his life, that the meeting in Stockholm was important, because it was there that plans were made to work for the establishment of geophysical institutes both in Bergen and in Tromsø, as soon as it was possible. A geophysical institute in Tromsø was to have the weather forecasting service as its main task, and secondly terrestrial magnetism.[280] Although these plans were unlikely to be executed, as long as there was still a war going on.

With them on the visit to Stockholm was, Bjørn Helland Hansen (1877-1957). He had been with Birkeland on the dramatic journey to Haldde when they were taken by surprise by a snowstorm, and a number of his fingers froze and had to be amputated. Therefore he never became a doctor, but he had begun to work with Vilhelm Bjerknes, and was regarded as being a leading figure in Norwegian marine research.

During the stay in Stockholm the small group had also visited, the Hydrographic Bureau. Here they saw among other things some registrations of remarkable oscillations in the level of water in lake Vättern, presumably generated by the wind. This inspired Ole to think about the probability that storms in the sea could be generated in a similar way. He wrote that he had talked to some fishermen earlier, who had informed him about how, "a heavy sea", that was when the sea swelled up, could occur even when the sea was as smooth as a mirror, which could be up until a day before the actual storm set in. In a letter to Birkeland he presents the idea about starting research of this phenomenon by registering such, "storm waves". The letter from Ole covers twelve pages as the idea is written in detail.[281]

At that point of time we know that Birkeland, was in Egypt and having problems with sleeping and homesickness, while at the same time the war was making it difficult to contact home. This situation ate away at Birkeland's state of mind. Even so he wrote an answer of encouragement and support to Ole dated Helouan 15th Feb. 1916.

> Dear Krogness. I have received your letter, newspaper report and dissertation on nature, and all of that has given me a great joy in this time of sorrow. I am sure, that over time, the Haldde observatory will receive a scientific halo over it, as your photographs show that it already has of the

[280] Devik1971:55.
[281] Letter from Ole to Birkeland 05.08.1915. Alta Museum archive.

Northern Lights. There are many discoveries of great significance that are in the making.

Ole in the barber's chair, Archive: National Library of Norway - Photo collection.

He is positive to Ole's new plans. "The plans are extremely good [...]. They are so good that they will become a reality." He mentions Bergen and thinks that one of the faculties there, "with especial reference to the polar region, geophysics and nature, yes that would be brilliant." Even so it was evident that he has some doubts with regard to Haldde observatory's fate. "[...] it was of course thoughts about Haldde observatory that brought about these uncertainties". Therefore he gives Ole some advice, "One must hold out as long as possible so that you and Devik can attain great things."[282] Moreover he had just (in October 1915) completed a dissertation that he had sent to a number of publishers, "so that it at least it

[282] Letter from Birkeland 15.02.1916. State archive, Tromsø, Haldde observatory, Folder 7.

should have a chance of getting in somewhere." It is one of his most important dissertations, he says, and has the title, "*Are the solar corpuscle-rays that penetrate into the earth's atmosphere negative or positive rays?*"

This looks as if Birkeland has found a solution to the problem Ole had presented to him, about the publishing of observations that resemble those done by Størmer and Vegard.

# CHAPTER THIRTEEN

# A CHILD IS BORN

A new child is born, Synnøve Louise Krogness. She was the first to be born on Haldde Mountain. Photo: Ole Andreas Krogness. Archive National Library, Photo collection.

*Yes, you were dismayed when you heard that the midwife had not come. She could have hurried a bit more, but she thought that I had estimated the date of birth incorrectly, even though I had given her the actual date.*[283]

On the home front the time for Ole and Dagny's second child's arrival is near. Family Krogness were back from a summer break in the south, to get things ready for yet another new year on the top of the mountain. Their first child, Oa, was born in Bossekop. This time Dagny knows what to

[283] Letter from Dagny to her mother-in-law, 03.10.1915.

expect, and has decided to give birth at home on Haldde. The birth will then not delay any research work, as it had when Dagny moved to Bossekop. Research and the everyday life can then exist side by side and nothing can take away any attention from what was about to happen, or stop the forces that would soon be set into motion.

Dagny has arranged with the midwife that she is to come up to Haldde a few days previous to the delivery date, and to be on the safe-side Dagny's mother has come to help. Dagny had herself planned the last part of her journey to the house on Haldde, both with transport and time, so that everything should go as normal as possible and not cause an early birth.

> Although I was very worried the journey up went well. Ole had borrowed a cariole, a two wheeled horse carriage, from the mountain's summer pastures, but as you know it bumped and jolted even so. I was quite dizzy when I arrived, but no pain. My mother sat in an armchair with springs on the way up the mountain, with Oa on her lap. She almost broke her hand when she was on the boat, but it had just come out of joint, that is the long finger had, but she got a man to pull it back into place. She suffers with terrible pains some hours during the night. Now it is swollen, but we hope it will improve [...]. Today we have emptied the suitcases with my mother's help, and soon I shall put all the baby clothes ready so they are in order. Miss Haga is to come 3rd or 4th September. Hope it will be in time.[284]

Luckily despite a little discomfort, everything is all right with the pregnant Dagny when they arrive at the house. Dagny's mother installs herself, and is ready to help during the last month of the pregnancy. The first time she gave birth Dagny had felt that she was a long way from her family, but this time her mother is there to help when the baby arrives. It is mainly for moral support, as her mother has said she will have nothing to do with the actual birth. She says, she has never helped with a birth before, and she does not want to start now. "My mother had ten children herself, but had never seen a baby born, but I had when I was 14 years old."[285] Dagny had been on a visit to her aunt and uncle, and helped the midwife when their daughter was born. "I thought it was wonderful to see a newly born child with all its limbs complete, and just everything."

---

[284] Letter from Dagny to her mother-in-law, 10.08.1915.

[285] Memoirs for grandchildren, written by Dagny, 1976.

Mother Guldberg together with her daughter and son-in-law, and some visitors. Photo: Ole Andreas Krogness. Archive: Private, Mette Bunting.

Her mother-in-law has also, like the previous time, sent some cloths and other things that will be needed before and after the birth. Ole has made sure that the telephone line is open all day, so that they can ring to the local doctor for advice and help if necessary. Dagny needs to know that everything is in place, and that the midwife arrives as arranged, in accordance with the agreement she has made with her.

> I am so pleased that I got Miss Haga. I have heard that she is very clever, so that is good. But she is not coming before the first day of September.[286]

It was smart of Dagny to have everything ready for the birth in good time. The statistics for that time, showed a high level of stillbirths and child mortality in Finnmark. If we count the number of stillborn and children who died before they were one year old, in the period 1911-1915, it was equivalent to every tenth child in the county. The family was very isolated on the top of Haldde, but not so much that they were unaware of the problems of a newborn baby's chance of living. Research of infant mortality had shown that survival depended on living conditions during

[286] Letter from Dagny to her mother-in-law, 19.08.1915.

pregnancy and the first year, as well as supportive help during the birth.[287] Dagny has correctly chosen qualified birth help as well as a good diet both during and after her pregnancy. At that time in Finnmark few pregnant women had access to a midwife, but often had experienced neighbouring wives, who came when needed.[288]

Little Oa dressed in traditional Sami clothes, a "*pesk*" and "*skaller*", a Sami jacket and footwear, made of reindeer skin, to stop him freezing in the Arctic winter. Archive: The World Heritage Rock-Art Centre, Alta Museum

You would have thought that everything is in place, for a tranquil and good birth, but living on the top of a mountain often results in plans not quite turning out as one expected. Dagny's description in later years, tells us that even with all the preparations the birth is dramatic. Despite the agreement with the midwife, she is not there when the birth begins, as it is a bit earlier than expected. In her memoirs for her grandchildren Dagny wrote, "Synnøve decided to come two days early, so both my mother and

[287] Statistics Norway1966, Bore 2007:45 and Elstad 2006:51ff.
[288] Elstad 2006:63ff.

your grandfather had to help deliver the baby."[289] Whether they try to get hold of the midwife by telephone is not known. But in one way or another they find out that the telephone is not working, and the first assistant Hoel takes responsibility. "Hoel ran to Kåfjord when they found that they had forgotten to connect the telephone."[290] In the house Dagny's mother and Ole have no other choice than to help Dagny, with the birth. Ole takes command, and studies a medical book they have there, and to cut a long story short, he does a good job. A fine healthy child was born and all went well with her.

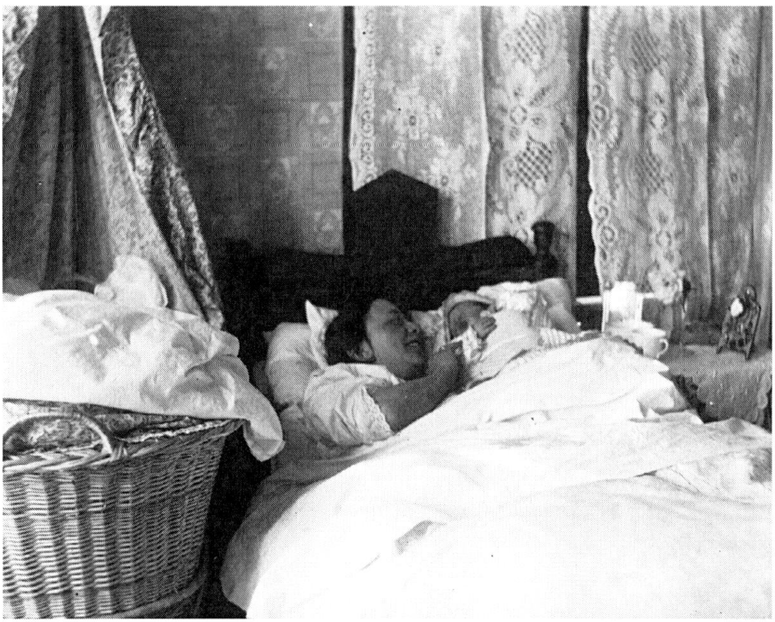

First photo of Mother and child. Photo: Ole Andreas Krognes. Archive: National Library, Photo collection.

Dagny helps and is completely ready. She describes an amusing detail in her memoirs, which tells us about the interaction between Dagny, her mother and Ole during the birth:

[289] Memoirs for grandchildren written by Dagny, 1976.
[290] Letter from Dagny to her mother-in-law, 14.09.1915.

> "[…] I asked if I should hold the lamp when Synnøve came. I was afraid that they would not cut the umbilical cord properly, but they did it perhaps too well, because the midwife had to do it again.[291] A fine healthy child was born, and all was well with her.

But the story did not end there. The placenta sat firmly fixed, and Dagny had a lot of pain:

> She had to take the placenta so I had just as much pain after the baby arrived as before the birth. I lay there two or three hours waiting for her (the midwife). Yes, mother will be able to tell about that memorable night. Mother was certainly more nervous than I was, I relied so much on her."[292]

An unqualified helper cannot do anything about removing a placenta that will not come out naturally. Therefore Ole and old Mrs Guldberg could not do any more to help Dagny. When eventually the midwife rang from Kåfjord to say she would arrive soon, it was Ole who took the telephone according to Dagny.

> "Yes, now the girl has also arrived", your grandfather said. Three hours later when the midwife came she apologised saying that, "such wives usually calculate incorrectly". My mother said to her, "Take off those clothes and put on an apron and wash your hands, because here you have to be quick." The placenta sat solidly in place and had to be peeled out, but all went well.[293]

Luckily all went well in the end with Dagny, despite the serious complications. Miss Haga's reputation as a good midwife, showed to be correct. So, with that, all was well in the world's most northerly Northern Lights' observatory, both with mother and daughter, Synnøve Louise.

> Yes, here I am lying in bed, I will be getting up the day after tomorrow. I am really fine, am waited upon and looked after, the little sister is fairly good natured, Oa is crazily enthusiastic about her, I have become nothing now that his sister has arrived.[294]

---

[291] Memoirs for grandchildren written by Dagny, 1976.
[292] Letter from Dagny to her mother-in-law, 03.10.1915.
[293] Memoirs for her grandchildren written by Dagny, 1976.
[294] Letter from Dagny to her mother-in-law, 14.09.1915.

Dagny writes this to her mother-in-law. Presumably she means that she herself has suddenly become of less interest for Oa, now his little sister has all of his attention.

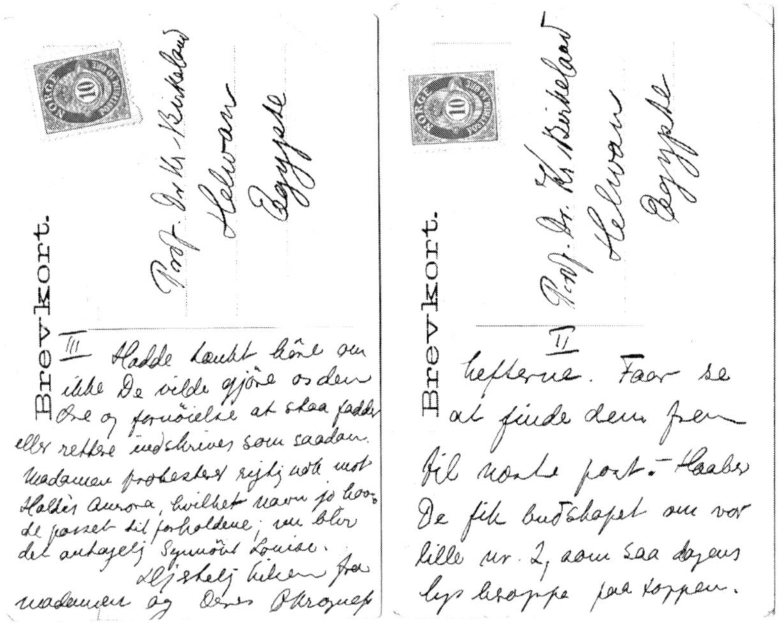


Postcard to Professor Birkeland in which Ole asked him if he would do them the honour of being Synnøve's godfather. He reveals Dagny's resistance to call the child Haldis Aurora. Archive: Observatory on Haldde's archive in the State Archive in Tromsø.

It appears that the birth on the mountain was dually noticed down in the village. "I received quite a lot of after-birth food, sour cream porridge from Heitmann and three litre of cream from Magda's sister, flowers, and a fish pudding from Mrs Feyling."[295]

This is another kind of description than the one Dagny gave after the first birth, when she had no family member there, which was hard for her to bear. Now they know quite a few people in the local milieu, something she is aware of, and which she appreciates.

[295] Letter from Dagny to her mother-in-law, 03.10.1915.

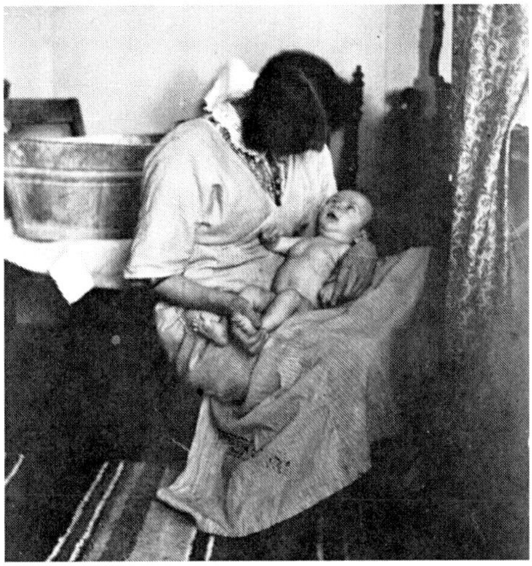

Time for a cuddle after the evening bath. Archive: National Library, Photo collection.

First family portrait with the new baby. From the left: Dagny, Oa, Ole with Synnøve. Photo: Ole Andreas Krogness. Archive: National Library, Photo collection.

# Chapter Fourteen

# Progress in the Research Colony

Having a chat outside, about research, in the good weather. Archive: National Library of Norway - Photo collection.

*It is such a serious time for Norway just now, and it is possible that we will not be spared from the war.*[296]

A short time after Synnøve was born, Olaf Devik and his wife, the other Dagny, arrived and moved to Haldde. On the tour up Olaf had put on skis and had gone all the way up on them, while the others travelled with the observatory's permanent driver Johan Olsen, who had fetched them from Heitmann's house.

"The castle, like a thieves den, from the age of chivalry, hung high up on a mountain peak, is what Haldde observatory looks like when one

[296] Letter from Dagny to her mother-in-law, 04.08.1915.

comes over the mountain ridge from Kåfjord, and can suddenly see it." This is how Devik describes his experience as he draws nearer to his new home. He also describes how extremely warmly they are welcomed by the whole colony, which was composed of Dagny and Ole, the little two-year old Oa and the newly born Synnøve. In addition to this, the assistant Hoel and the Lukkassen family, but the maid is not mentioned. Dagny's mother was still there in connection with the birth. Ole plays a welcome march and the party began, both for the newcomers, and for the first child born on the top of this mountain.[297]

The view, as seen, from the veranda on the second floor of the main house. The veranda was used to register observations. Photo: Ole Andreas Krogness. Archive: National Library of Norway - Photo collection.

During the journey up along the coast, Olaf had a longer stay in Mo i Rana, where he made a valuable contact with Hans A. Meyer, the head of the firm L.A. Meyer. They had discussed both weather forecasting problems, and cultural questions about Northern Norway. Later Meyer was an important person in the planning of how the Haldde activities should be continued in the future.

[297] See Devik 1971:57-58.

Surprisingly, Sandstrøm came to Hammerfest about the same time with Olaf and Dagny Devik. It appeared that he was also going to Haldde. His mission was to discuss with Ole the experiences he had made with the study of hurricanes, and to take a closer look at the special observatory high up on a mountain. With him, Sandstrøm had a new instrument to register extremely long sea waves. Clearly Ole's idea about the significance of waves as an element in predicting the weather had obviously awoken his interest. The instrument was expensive and Sandstrøm had paid for it himself, so it had been a spartan journey for him on the coastal liner up north. To Devik he commented that he, "had found out that water was actually drinkable!"[298]

With new residents, and visiting guests it was even more crowded on Haldde. Dagny is occupied with her children; and she especially mentions the newly born baby, in her letter to her mother-in-law. Her own mother has just returned to Kristiania.

> My mother left here yesterday, which was sad for me, but the baby managed to start screaming from the minute she left, and has made a mess of herself, so that I was kept busy for a while. Ole and Sandstrøm left with mother, as they were going to Bossekop, in connection with their research.[299]

In the late autumn it appears that Ole goes on another journey. It is in connection with a meeting or a conference, where he is expected to give a lecture, which has been rather delayed. Ole has prepared himself, and it is decided that he should take part, but Vegard was also at the meeting, and it was Vegard's and another lecturer's turn to go first. This is made clear in a letter from Dagny to her mother-in-law in December.

> I wonder if Ole will manage to come up here for Christmas, if he is able to. Thank you for your last letter, I am sorry that I made such a lot of fuss in connection with Ole coming the 19th December, but he was decisive about this, even so it turned out that Vegard and another person held lectures first so that Ole had to wait.[300]

A delay for Ole also means that his journey home will be hindered, and the connection between the south and north of the country was not especially quick. The post from Kristiania to Kåfjord and up to the house took normally six to seven days. When the local boat from Hammerfest to

---

298 Devik 1971:57.
299 Letter from Dagny to her mother-in-law, 03.10.1915.
300 Letter from Dagny to her mother-in-law, 11.12.1915.

Kåfjord was later restricted during the war, to once a week, the post took up to ten days. Devik explained that he and Ole took it in turns to travel south to the Ministry of Church and Education to confer about their work, and plans if there was something they were extremely interested in. It turned out that this was worthwhile. "And we never complained!"[301]

Dagny and Olaf Devik. They came as a young and newly married couple to Haldde. Archive: Private, Mette Bunting.

It is quite clear from the letters that there was more humour in the colony after more people came to live there. The researchers seem to have found a humorous tone in their scientific work, and this also influences the everyday activities. In a letter to Ole's mother Dagny and Ole take it in turns to inform her:

> Dagny: "It is pleasant having the Deviks here during the winter, we get on very well." Ole: "Dagny should have been my secretary, but I must also add a greeting and congratulations." Dagny: "Ole is constantly interrupted by the telephone, but does not give up." Ole: "It is the electricity board that have decided to make a decision on your birthday. Hope they succeed so that we can get an electric bed warmer, cradle, travel machine etc. Yes there are many plans in the air up here too. Apropos, fire. We woke up here today to a dull thud as a self-made damp machine in the cellar exploded. Some people had quite a shock, but others thought it was funny.

[301] Devik 1971:59.

Our primus went completely west, and madam was extremely affronted, as it had been stolen without her knowing, but she will soon get over that."[302]

Life with research and everyday life are going favourably well. The men do research. The women look after the house and children, everything is going merrily along. However not even the researchers are without daily pursuits. Ole is working hard to try and get electricity up to the mountain top. With small children in the house there are a lot of nappies to wash, and electricity was important for melting snow into water. Eventually they found a solution in Kåfjord's hydropower station. It was this station that had been responsible for providing the mining industry with electricity, but when the mines were closed, there was some spare capacity, and they were trying to use it for the people in Kåfjord and on Haldde. Through long preparation, calculations, letter writing and fundraising over many years. In 1917 they succeeded in installing electricity on Haldde and the house in Kåfjord. This meant that the houses could be kept warm, also when the wind blew and when there was a lot of carbon monoxide.

The Electricity-Stock Company A.E.G in Trondheim was extremely helpful in getting the necessary equipment for the electrician, who was to do the work. The electrician demanded a double wage, based on how difficult the task was, but the company refused this. What they offered was to cover the cost of extra winter clothes for him, as he was to begin in October, when there was full winter in Kåfjord. Through demanding hard work, iron cables were laid up the mountainside from the old copper works. Electric heaters were sent from Trondheim to melt the snow. By Christmas the electric current had almost completely been installed.[303]

Work and planning took up most of the researchers' time. They thought it was also an advantage that newspapers only came once or twice a week, because it meant they had a chance to read a book. Devik had met Eivind Berggrav before he travelled north, and complained to him that there was not a particularly well-equipped library on Haldde. Berggrav, who was the editor of the periodical concerning church and cultural matters, "*Kirke og Kultur*", and later the Bishop of Hålogaland, offered to lend Haldde colony the books they would like to read. It was not just a few parcels of books, but many, that were sent up to them from the capital later.

---

[302] Letter from Ole to his mother, 29.02.1916.

[303] Letter from Electricity-company in Trondheim, 6.10. &22.12.1916.

Lukkassen with the "instrument box" Archive: National Library of Norway – photo collection.

The new activities also involved more correspondence, for example a letter was sent 20th January 1916 from Haldde, to Fridtjof Nansen. Olaf Devik said thank-you for the reading material. It was the magazine, "*Spitzbergen natur*", which the postman had brought. He also thanked Nansen for allowing the Haldde researchers to use his instrument maker. "The boy appears to be clever, and we would like to keep him", but that had to be postponed as he had to do his military service. The contact with Nansen was of great scientific significance. Devik wrote, "We on Haldde touch upon oceanography, but in due time a thermodynamic interaction will develop between the sea and air, and come closer into our lives."

The research often required Ole and Olaf to travel. Olaf had to go to Røst in Lofoten to establish one of the first meteorological stations. He left, taking the local boat along the coast, and carrying his instrument box. It took about a week each way. Meanwhile Ole travelled to Bergen to hold a lecture in the, "*Muséforening*", the Museum Association, about thc work at Haldde observatory. At the same time plans were also discussed for a geophysical institute in Bergen. It was decided to start a collection towards funding this.

While the two researchers were away, the daily life on Haldde carries on as usual, with the wives, assistants, children and maids. Not everything was easy, even though there were more people living in the colony.

Strong and clear Northern Lights. Archive: National Library of Norway - Photo collection.

In a letter to her mother-in-law in autumn 1916, Dagny tells her that she is exhausted, and worried about many things, the children's health, a maid who gave in her notice, the war, Mrs Devik's struggle, and an assistant's wife who has continuous miscarriages. The children are ill. Oa has earache; Synnøve has choking attacks, which are extremely grim when she has them. Dagny herself has far too little time for the children. Oa is wild and uncontrollable. He has to be inside because of the bad weather. There is little help to get from the maid, who is only 15 years old. Dagny still wants to move south, but is afraid of the war reaching Norway, and especially in the southern regions of the country. An almost panic letter goes from Haldde 27th November 1916:

> "Dear Ole! I am so afraid that you will not come home before there is a war, and there is not even a telegram from you so I do not know where you are. I do not dare to travel to Tromsø either."

But Ole comes safely back from his research expedition, and the war still avoids them; just fear is left hanging in the air.

# CHAPTER FIFTEEN

# MORE BIRTHS

Dagny Devik with her first-born, Finn. Photo: Ole Andreas Krogness. Archive National Library of Norway – photo collection.

*We have quite nice weather, the ground is still bare and there is quite a lot of rain so that we have water for the food and Søssa's nappies and dresses, she makes such a mess of them now that she is crawling.*[304]

Synnøve Louise was not the only child to be born on Haldde. Dagny Devik gave birth to two children during the three years she and Olaf lived there. Finn was born 5th May 1916, and Ingebjørg 28th January 1918. All

[304] Letter from Dagny to her mother-in-law. 08.09.1916.

the births there went well, although each of them had their special circumstances and stories.

When it was Mrs Devik's turn to give birth, she has also prepared things as well as possible. The midwife was there in good time for the delivery of her first child. That was, by the way thanks to their husbands, something which both women write about. Ole and Olaf found out something about the moon and the delivery time for the births, and when Dagny Devik is expecting her second child, the men say the midwife has to be there in good time.[305] Their calculations result in the midwife arriving two weeks before the birth starts, which is of course totally unaffected by the moon's course and the scientists' calculations.

The midwife, Mrs Fiske, has many amusing stories to tell during the time she has to wait for the birth, stories that make the women laugh until they cry. The waiting period is a welcome break for all of them, with much gossip over coffee. The stories Mrs Fiske entertains them with, are from the time when she was doing her midwifery training at the old clinic for women, at the National Hospital in Kristiania, "*Rikshospitalet*". One of the stories she tells is about the notorious "*Emperor*", the legendary Chief Physician Schönberg, who had the reputation for not being so happy about anaesthetizing mothers, who were about to give birth.[306] He was, from what one understands, a determined man with distinct opinions, and was on one occasion appalled over a woman in labour complaining about being served pea soup. Mrs Fiske meant that you could not offer a woman in childbirth such food, because it would give the baby unnecessary air pains after drinking the mother's milk, but for the chief physician it was just nonsense. The baby will not have any pain from the mother eating a little bit of pea soup. Mrs Fiske talks non-stop, describing and embroidering to such an extent about all kinds of things, that they break down in laughter.

The midwife has not forgotten why she is there, so that while she is entertaining the wives, she has made a personal programme for Mrs Devik. During and between the stories Mrs Devik has to trot through the house to encourage the birth to start, which does eventually happen and everything goes well for both mother and son.

Not everything concerning health and pregnancy went quite so harmoniously and uncomplicated there, as it had done for the two Dagnys. Mrs Hoel had a more complicated case history than the other two women. Mrs Devik described her like this, "[…] she was a bit hysterical and quite

---

[305] Memoirs for her grandchildren written by Dagny, 1976.

[306] Conversation with grandmother 27.08.1985. Dagny Devik and her children, Ingebjørg, Dagmar, Ole.

crazy about illness. So there was nothing we could say, and the others also had to keep quiet."[307] It was not quite surprising that Mrs Hoel was a "little hysterical", when they talked about hospitals and illness, if we take into consideration what she had been through in her young life. She had nearly lost her life when she had a really serious case of appendicitis, and in addition had experienced a number of spontaneous miscarriages while she was on Haldde. It is not difficult to understand that to be pregnant and lose the foetus every time, led to a deep sorrow. One can only imagine what it must have been like to be surrounded by happy pregnant and birth bearing women, in such a small place, when you yourself could not manage to carry and deliver a child. Dagny helped her when it was necessary.

> Today we are arranging for the doctor to come and see Mrs Hoel. She has had another miscarriage. I do not understand how she is created. Luckily I was able to help her, but it was the most unpleasant work I have done. I would much rather have helped with the birth of a baby.[308]

Mrs Devik gives birth once again in 1918. They hope that this time it will not take too long to wait with the midwife. The same midwife, Mrs Fiske, comes. She soon finds her story-telling form again. This time the labour pains start quickly and Mrs Devik is put into the bedroom. While they are waiting, the women drink coffee and play cards, and enjoy each other's company. Mrs Devik is doing well in the bedroom, but says that now she thinks it is time for the birth. In accordance with the midwife, she shouts out, "I think the baby is coming now." The midwife on the other hand takes things calmly and says," Oh, nowhere near."[309] So Mrs Fiske continues to enjoy the coffee that Devik's brother had sent them all the way from Rhodesia. She does not go in now and again to check, and finally Mrs Devik is completely quiet in the bedroom.

---

[307] Conversation with grandmother 27.08.1985. Dagny Devik and her children, Ingebjørg, Dagmar, Ole.

[308] Letter from Dagny to her mother-in-law, 26.11.1917.

[309] Memoirs for her grandchildren, written by Dagny Devik, 1976. The same information can be found in, "*Samtale med farmor*", Conversation with grandmother 27.08.1985. Dagny Devik and her children, Ingebjørg, Dagmar, and Ole.

Oa loves to play with the older Lukkassen girls, shown here on the small piece of ground outside the house. Photo: Ole Andreas Krogness. Archive: National Library of Norway – photo collection.

When Mrs Devik announces a bit later that the baby has arrived, Mrs Fiske jumps up and runs in to help. Her comment is that Mrs Devik has not yelled enough or made enough fuss, and it is completely unheard of to be so calm, as she had been. The last child to be born on Haldde has arrived, a fine and lovely girl, Ingebjørg.

Mrs Devik tells her children some years later, how much flowers meant to her when she lived on the top of that barren mountain. It meant a lot to her when she, as a newly fledged mother, received thoughtful consideration from the village. Right after the birth Mrs Heitmann arranged for an unusual present for her:

It was impossible to use a pram when it was snowing, but a kind of sledge was needed, which we can see here, and Synnøve was packed down into it. Ole and Mary play in the background. Archive: World Heritage Rock Art Centre - Alta Museum

Olaf Devik proudly holds his first born, Finn, who is packed into a "komse" a Sami cradle. Photo Archive: World Heritage Rock Art Centre - Alta Museum

> Mrs Heitmann, who was motherly, round and kind, had lots of children. A pair of galoshes, old galoshes, suddenly arrived, and between the two galoshes were some crocus, which had just opened, grown from bulbs that she had nurtured. Can you understand what that meant to me? Wasn't that lovely! I can never forget that.[310]

The story about the midwife gives us some insight into what it was like to work as a midwife in the local community, when those about to give birth chose to have a home birth, and the home was high up on a mountain. In her memoirs, Dagny Krogness makes a comment about Synnøve's birth, which also tells us how she interpreted the midwife's interest in her profession, "That midwife was not interested in her profession. She said that here in the north we cannot be anything, because we have no schools. The only possibility is to be a midwife."[311] What the midwife meant by this we do not know. It is possible that at times the midwife thinks that she would have liked another profession rather than being there or in other similar situations, when she has to go out in all kinds of weather, and to all kinds of establishments in difficult inaccessible places. Perhaps she would have taken another education, if it had been possible, but she could only be a midwife because in that region there were no other schools.

From her lovely garden Julie Heitmann picked flowers and sent them up to the ladies on the mountain. Archive: Private, Frithjof Heitmann.

---

[310] Conversation with grandmother 27.08.1985. Dagny Devik and her children, Ingebjørg, Dagmar, and Ole.

[311] Memoirs for grandchildren, written by Dagny, 1976.

# CHAPTER SIXTEEN

# CHILDHOOD MEMORIES OF HALDDE

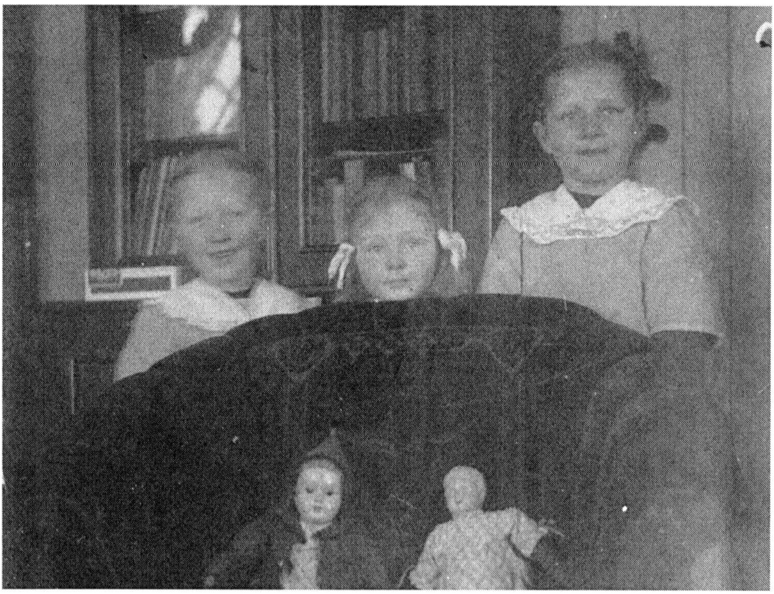

The three Lukkassen sisters. The top of the mountain became a home for a whole flock of children. Archive: The World Heritage Rock-Art Centre-Alta Museum.

*Yesterday the cow arrived so Magda has enough to do. Mrs Lukkas has not arrived yet. It is good for me now that I have some milk, because I do not know what else I should drink.*[312]

As a result of Ole's application to have a general caretaker on Haldde, the Lukkassen family had moved up to the small colony from Kåfjord in the spring of 1914, just as Ole told Birkeland in his letter. The family

[312] Letter from Dagny to her mother-in-law 03.10.1915.

Lukkassen included a mother, father and three daughters; Aagot, 8 years old, Borghild 6 years old and Mary 4 years old.[313] That was the beginning of the most populated period on Haldde.

Since the Lukkassen family had a cow with them, Dagny had a solution for the irregular milk supply. That was how cattle farming became part of their practical work. The cow was placed in the stall where Ole had kept a horse the previous summer. In her memoirs Dagny describes the arrival of the cow:

> We decided one winter to have a cow up there, and up came a cow. It was put on a sledge and taken up during winter conditions. But the cow was expecting a calf, and one evening it was delivered. I had gone to bed and was lying down reading Hanna Wisnes' book, and then Devik came into Ole and told him the cow had calved. Then we all had a busy time. Devik made a bin, Ole looked after the cow and calf and Dagny and I made a warm drink for the cow, and we made sandwiches, and we had to have coffee for such a special occasion. But we slaughtered the calf in the spring, and Ole and Olaf wrote a song of mourning. I thought it was horrible eating the meat.[314]

As they now had their own cow on Haldde, there was no fear of them having milk from a cow with tuberculosis, and they always had milk, even if the weather allowed for supplies to be transported up to them or not.

Our material contains a unique interview from 1991 with two women, who were children when they lived on Haldde, the sisters Mary and Borghild Lukkassen. At the time of the interview Mary was over 80 years of age, but gave quick responses and very reflected answers. Her memories from her childhood on the mountain gave us some insight into the everyday up there, from a child's point of view; even though she and her sister said that they could not remember that much, as it was so many years since then. Below we tell her story as closely as possible to the way she herself told it in her own words when she was interviewed.

Her father was appointed by Krogness as a caretaker, with the main task of operating the winch that brought goods up the last steep slope to the house. Actually, their father's sister Magda Lukkassen had already worked some years for the Krogness couple, so that for the Lukkassen girls it was like moving up to their aunt on the mountain. The family had a cow with them. It was important that the children should have fresh milk. Mary remembered that the cow delivered a calf while they were living on Haldde. They had also a goat with them, which they put into the barn on

---

[313] Nielsen & Petterson 1999:37.

[314] Memoirs for grandchildren, written by Dagny, 1976.

the mountain. That was mainly so that there was milk for the adult's coffee. The goat was given the appropriate name, "*Kaffimelka*", (Coffee milk). The girls' mother was mainly in charge of the animals.[315]

Mary said that when the girls got to school age they went to school in Kåfjord, and their mother lived with them in a house down in the village. The school was divided into two groups. By attending both groups every day she had a double amount of schooling, and was then able to live some of the weeks in-between, at home on Haldde. There was no talk of going home every day, as it was far too long a distance.

Mary liked to go skiing on her own. She did not want to ski back down to Kåfjord with the postman, when he delivered the post to the colony. It was much more fun to ski down alone without any adults. Once she had managed to leave before the postman arrived, and she flew away at a great speed down the mountainside, not a dangerous situation according to her. It was just the adults, who thought it was. If it went really fast she sat on her ski poles. Her aunt had heard some boys discussing Mary's skiing skill. "She is clever," they said, "but she has a bad technique."

She remembers a dangerous situation when her father was to carry out the observations on the top of Talvik Mountain, which was 3–4 kilometres from Haldde.[316] Fog came and the snow drove really hard, and it was so thick that it was impossible to see anything, and her father lost his way. He kept going but just went round in circles, he was hungry and tired, but did not dare to sit down. Finally he saw a small wood and some scrub bushes, and realised that he was getting closer to a village. It turned out to be Talvik. Some people from Haldde had been out looking for him, and found that an avalanche had covered his ski tracks. They summoned a search party from the village. Whether they found her father, or her father found his own way to the village, she did not know. Quite by chance the local boat that evening called in at both Talvik and Kåfjord, so that their father got home. Luckily the search party could also return to the village. You can imagine the surprise of the Lukassen family, who were at that time in Kåfjord, when the door opened and the girls' father came in, as none of them had known that he had disappeared. She remembers:

---

[315] We see that Dagny's memoirs about the arrival of the cow (See Chap. 13 – New Plans) is not quite identical with what Mary tells this is due to memories becoming rather old over time. The authentic quotes from letters are more credible.

[316] Devik 1971:32.

Mary Lukkassen, the youngest girl. Archive: The World Heritage Rock-Art Centre, Alta Museum

> […] the fright of seeing father, and that he had got the observatory apparatus on him, made us realise that something had gone wrong. When we got to hear about the hardship he had been through, we were all sure that God had kept him safe, and given him strength to manage the long and difficult journey down the mountain.

There was a good camaraderie between the Lukkassen children and the children in the village. Mary told them that before the Haldde children began in school, they went down to the village every spring and stayed there until the end of the summer, because when there was no snow on the mountain, there was no water either.

The weather was hard on the mountain. She remembers how the house shook even though it was built of big stone blocks. When there was really bad weather they could not go outside. Then the underground tunnels had to be used to get to the outhouse, barn, water-storage tank or the observatory. The snow was melted for water, and coal and wood was used for heating. There was also a type of wind, which they called, "exhaust" wind, she says. When the exhaust wind was there it was no use trying to light the stove, because the smoke blew in. She thinks they only had electric lights when they lived down in the village, because there was a light generator for their use in Kåfjord. The house, in the village, also functioned as a thoroughfare for people who came and went to the colony on the mountain.

The Lukkassen family house in Kåfjord. Photo: Ole Andreas Krogness. Archive: National Library of Norway – photo collection.

Mary stresses that the men's work on Haldde was hard. Night after night they stood outside to collect information about the Northern Lights. It was a struggle. Her father stood on the top of the mountain, most of the time, with the large searchlight that they used during the Polar nights. She thinks it must have been a strain for him, even though he was robust and used to the climate in the north.

The sister's father and aunt Magda, who was the maid on Haldde, plus another relative, who was visiting them that day. Photo: Ole Andreas Krogness. Archive: National Library of Norway – photo collection.

In connection with the musical life on Haldde, Mary remembers that Krogness played the piano when they got together to enjoy themselves. Her mother played the guitar and sang for the children. Her father also sang for them, and her mother read to them from the school's reading book as well as from a children's book about Meg and her children, called, "*Lille Megs barn*".

The games they had on the mountain, for indoor use, were the ones that were normal for that time. There were dolls, horse and wagon, and building blocks and they played, domino and played with "*munkekuler*", using round stones or glass marbles. Mary had a lovely doll, which her mother sewed clothes for, but she never had a cradle, she says with a smile. On the other hand Borghild had a doll's bed. Outside there were sledges or they played in the snow, which was fun. But there were always adults there, because the mountain plateau on which the house stood was small, and the mountainside surrounding it was dangerously steep.

She cannot remember if there were any births up there, because, "at that time such matters were not discussed with children." Neither can she remember seeing a baby being bathed, but she assumes it was in a tub in the kitchen, something that was normal then.

She tells about parties and celebrations, for example Christmas, and her parents' ten years' wedding day, that was celebrated with the others, who lived there. The Lukkassen family often had visitors. Once a cousin from Komagnes visited them:

> We girls gave our father a big surprise. He wanted to go to bed, but we thought it was too early. We threatened to come and tip up his bed, "Yes, just do it," father said. When he had got comfortably into bed, the "strong young ladies struck again, and the bed was rolled over with much laughter. "I did not think that you were all that crazy", Father said.

There is no doubt that those who lived on Haldde thrived there. Some years earlier the sisters had visited Mrs Devik in the nursing home. She said the best time they had experienced was when they were on Haldde. When the Lukkassen family moved to Tromsø later, Mary did not thrive there, but longed to be back in her, "fairy–tale castle", which was what she called the house on Haldde. It was safe there, and there was a good atmosphere.

The adults were concerned about how to bring up their children. They discussed it but were not completely in agreement. Some thought children should be spanked and some disagreed. "I have been smacked", she says, laughing. Mary the youngest in the sister flock describes herself as wild, a bit naughty and someone who could not sit still. The eldest sister used to warn her and give her advice, to prevent her getting into trouble, "You know Mary that when mother says, no, she means no!"

The sisters tell, for instance, about the village, and that Heitmann owned a number of buildings and a shop. Their father worked there when the copper mines were working. This is how Mary concludes her remarks about Heitmann:

> Heitmann took over, you could say, most of Kåfjord after the mines closed. He became an important man. He owned most of Kåfjord. They say that someone in the village asked Heitmann, "Do you also own the lid over Kåfjord?" Kåfjord village is, what can I say, like a small casserole.

There was a big class difference in the region. There were many that were not very well off, and in some cases parents had to be separated from their children, and place them with other people. Therefore a children's home was built in Korsfjord and one in Kåfjord. Mary tells us that it was church people from down south, who came and looked after these children. They also started a Sunday school in the village. The Lukkassen sisters began there. The eldest sister was, prime motivator. She turned up with a whole flock of children behind her, to go to Sunday school.

Karen and Karl Lukkassen, the parents of the three small girls. Photo: Ole Andreas Krogness. Archive: National Library of Norway – photo collection.

The Lukkassen family were in Kåfjord from 1914–1918. In 1923 Karl Lukkassen was back on the top of the mountain, as an experienced assistant for the new Director, Hilding Köhler, but in 1925 the whole family moved to Tromsø. Karl was appointed as caretaker at the Geophysical Institute.[317] That his daughter Borghild got a job at the Meteorological

[317] Nielssen & Petterson 1993:37.

Institute, in Oslo, probably had something to do with her father's career as assistant in Northern Lights' research, and the weather forecasting service. All three Lukkassen girls settled in Oslo. Their parents followed them when their father retired.

# CHAPTER SEVENTEEN

# SOCIAL LIFE FOR CHILDREN AND ADULTS

Coffee break out in the open air for wives, maids and children. From the left (the adults) Dagny Krogness, Karen Lukkassen, Magda Lukkassen, Sanna Pedersen and Dagny Devik. Photo: Ole Andreas Krogness. Archive: National Library of Norway, Photo collection.

*"Tomorrow the maid will be preparing the fish, we shall have the residents for dinner the 14$^{th}$, and fish is the easiest and cheapest."*[318]

Most of the time there were four families on Haldde, each of them with their own housekeeping. The families consisted of seven children, four couples and two maids. Dagny no longer had a special responsibility for a large household, and could therefore concentrate on being wife, mother and neighbour.

[318] Letter from Dagny to her mother-in-law, 10.11.1917.

Complaints about loneliness and lack of people to talk to are now replaced in letters with descriptions of the children, domestic life and social happenings. Already in January 1915, when Olaf Devik visits them, Dagny realized that there would be changes. She writes to her mother-in-law:

> I am looking forward to winter if the Deviks' are coming, because you have no idea how little company there is with Hoel, [...] but even so, one must make it as comfortable as one can. I think it is amusing and exciting to exchange thoughts and to discuss things, but there is never time, so that when Devik came, it was as if one looked into another world, and one must involuntarily think, how much richer everything would be, but everyone cannot be equally intelligent.[319]

Up until Oa was two years of age, he only had adults round him. Therefore Dagny and Ole were pleased that the Lukkassen family, who would be the first family to move up there, had children. Ole describes the change. "There has been a bit more life now with Mrs Lukkas and the three small girls. The boy has not completely learned how to play with small girls yet, so things are not going as well as they should, it seems."[320] Despite this, it was obvious that Oa needed playmates, even though he would rather be with his mother. Ole continues, "It seems he still prefers to be with his mother. He is extremely charming, but talking does not appear to be going so well, on the other hand he screams and fills his pants all day."

When the Devik couple arrive in the autumn of 1915. They are made welcome before they reach the house, just like others that visit the top of Haldde. Coffee and sandwiches are sent down on the, "kibben",[321] to the foot of Haldde Mountain, "because that was always the custom when anyone came from Kåfjord after travelling for a number of hours. One needs to take a breathing space for a bit, before climbing the last steep ascent up to the observatory."[322]

---

[319] Letter from Dagny to her mother-in-law, 07.11.1914.
[320] Letter from Ole to his mother, 16.11.1914.
[321] The simple cableway used to take wares up to the top of the mountain.
[322] Devik 1971:58.

It is hard for Oa to sit still for the photographer. From the left: Oa and Dagny with the sisters Aagot, Mary and Borghild Lukkassen. Photo: Ole Andreas Krogness. Archive: National Library of Norway, Photo collection.

In order to give the Devik couple time to settle in, Dagny invites them to dinner the first few days. The two wives get on well together very quickly, even though they have never met each other before, and they gradually spend more time together and with each other's children. At times, Dagny needs help in coping with her household, and two small children. Then Dagny Devik steps in and helps, especially when the men are away. "Mrs Devik helps me with the children at times, she has also slept in Ole's bed for two nights when Magda was in Kåfjord, and now we have cleaned all the rooms, yes, but not exactly spring cleaning."[323]

A deep friendship develops between the two couples. Living so closely together one would think that it would be a strain on a relationship, maybe especially for women, who had to get to know each other. It was also seldom they were anywhere else, except on the top of the mountain, enclosed in the house for many weeks at a time when the weather was bad. This seems to have strengthened the friendship of the two Dagny's. Mrs Devik tells about how harmonious she experienced her friendship with Dagny:

[323] Letter from Dagny to her mother-in-law, 11.12.1915.

> Dagny Krogness and I have never fallen out. […] We lived closely together. I think that we thought that as the men were so haphazard we ought to tease them, but we never managed to get them angry. This has, I am sure, meant a lot; they were peaceful people both of them. Yes it is true, that now and again one could be annoyed, but it was impossible to tease them. They simply could not be teased.[324]

Good humour and smiles in the snow and wind. Archive: World Heritage Rock Art Centre - Alta Museum

In contrast to their husbands these two friends never dropped the formal Norwegian address of, "De", as long as they were on Haldde. It is only when they move down from the top of the mountain that they use the informal form, "du", to each other. On the mountain they use," De", or use their title plus their surname; for example, Mrs Krogness and Mrs Devik. We also see this throughout Dagny's letters. Society's conventions are used, and they keep a formal tone, despite a life that demands more closeness, a great deal of each day, than is usual. Maybe the distancing through language is the key to them developing a harmonious friendship, when they live so closely together? Another thing that may have been of

[324] "*Samtale med farmor*", (conversations with grandmother), Dagny Devik, 27.08.1985, and her children, Ingebjørg, Dagmar, and Ole.

help is that they have their own flat, each with its own kitchen, which means that it was possible to withdraw and work undisturbed in their own part of the house, when they want to. Dagny Devik was a most efficient housewife, and maybe it was a good thing that she and Dagny did not have to share a kitchen. The friendship between the two families grew, and it would especially be an important support for Dagny Krogness in her life after Haldde.

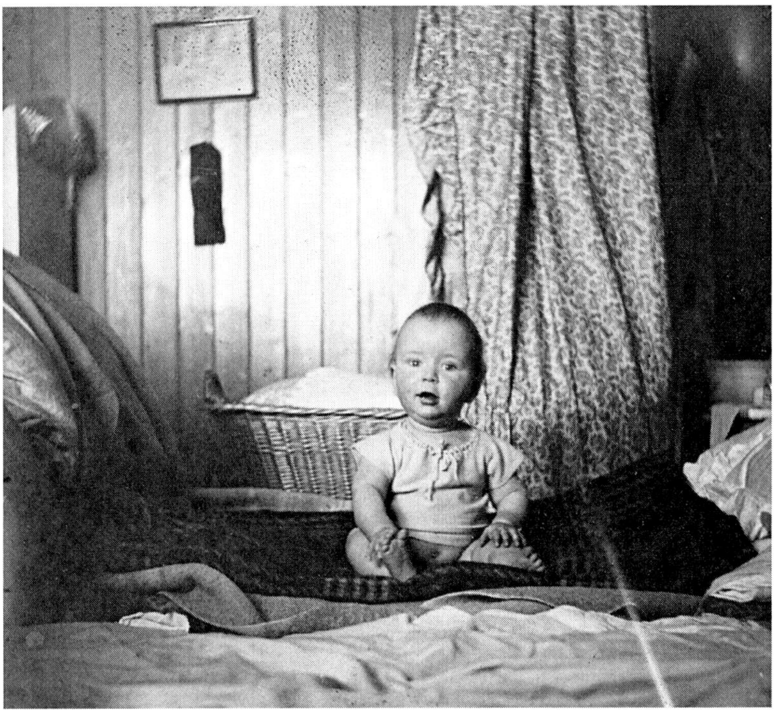

Oa sitting on Dagny and Ole's bed. Even though it was cold at times, there were also days when one could wear less clothes. Archive: National Library of Norway – photo collection.

With so many people in a small place there was naturally quite a lot of activity. The house is large, but just imagine all the weeks when the weather did not allow any of them to go outside, and then it could at times feel small and confined. Olaf Devik writes in his book,[325] that life there

[325] Devik 1971:51.

was extra tough for the women and children, when winter came and they were often cut off without any contact to the world around them. He believes that taking the extreme weather conditions into consideration, it was quite an achievement to build and run a place like Haldde. One gets an impression of the many challenges, such as when a hurricane blows round the top of the mountain, and snow runs like a river outside, dances at lightning speed, swirling round stone blocks in the scree's loose stones. While gusts of wind throw themselves at the house, so that the powerful grey stonewall shakes from the furious snowstorms on the mountains outside.[326]

To measure the wind they have to have special wind measuring equipment, as those used normally were not strong enough to stand up to the challenge. The fastest speed measured on Haldde was 65 metres per second, rather a lot, compared with the wind-power of hurricanes, calculated as being from 33 metres per second and up. The residents have their own scale for measuring the wind force:

> At 35 metres a second (35m/s) one could sail on the wind stream as if it was a river, at 40m/s one had to creep along, at 45m/s one had to slither one's way like an eel, but if it was 50m/s or more, one risked being blown away, unless one had something to hold onto.[327]

As a weather researcher, Olaf Devik loves a storm and when gusts of wind play havoc on the top of the mountain. For the women and children, it sets limits as to when they can be out and where they can go.

The women did not usually leave Haldde before the summer. The men, on the other hand, had some travelling to do, to look at weather stations and cooperate with other researchers in connection with their work. They could be away for quite a few weeks, yes, sometimes even months.

It was as Ole Devik pointed out, not always easy for the women when their husbands stayed away for a long time. Maybe it was especially difficult when war broke out. Dagny was uneasy when she did not hear from her husband for a while. In one of her letters she said that she needed to go to Tromsø, but felt she could not go, as it would involve extra work for Mrs Devik. Mrs Hoel had been sick again, and they had builders, who both worked and lived in the house. In one letter it appears that Mrs Devik has offered to have the children so that Dagny can go, but she is uncertain,

[326] Devik 1971:51.
[327] Devik 1971:60.

"[…] I do not dare to let her do too much, because then I will have her husband after me. I will go after Christmas."[328]

The women are ready for a ski tour despite grey weather. Photo: Ole Andreas Krogness. Archive: National Library of Norway, photo collection.

When one lives so isolated one becomes good at being happy about the small things in life. Mrs Devik was very fond of flowers, something it seems that their driver Johan Olsen knows about. She tells her children some years later, "[…] our driver used to pick some heather, and some dwarf birch twigs and such things, in the spring. And do you know that to pinch the birch, and smell it, oh it was so wonderful. And there was also bell heather."[329]

---

[328] Letter from Dagny to Ole 27.11.1916.

[329] "*Samtale med farmor*", (conversations with grandmother), Dagny Devik, 27.08.1985, and her children, Ingebjørg, Dagmar, grandmother and Ole.

Uneven ground on which to push the pram. Photo: Ole Andreas Krogness. Archive: National Library of Norway – photo collection.

They all made good use of the days when the weather was fine. They arrange ski races in places where it is flat enough to ski without being afraid of falling down the scree. Sometimes they place the dining table and chairs outside in mid-winter to have a picnic, "[…] held a ski race last Sunday, ski-hopping where everyone got a prize. Absolutely successful and will soon be repeated again."[330] Even though it was one of the darkest and perhaps coldest days of the year. Once the sun peeps out it is not long before some of them are sitting outside along the wall, where the sun shines. However, to take a tour out in the winter can have its challenges, as Dagny tells us:

[330] Letter from Dagny to her mother-in-law 12.02.1916 (*Comment from Ole at the end of the letter to his mother.*)

> One day I went skiing when it was good weather, but suddenly it started to snow or it was fog, and I was grateful that I found my way home again. We did not go skiing very much there, but if it was good weather we took the children for a short trip with the sledge, and had cocoa and sandwiches with us.[331]

The good weather had to be exploited. The wind could suddenly blow up. Dagny wants to let Oa sleep outside, as she is used to in southern Norway. She tells us in her memoirs that she had to take precautions. "When Ole was put outside to sleep in his pram I had to stand watch. I was afraid that a wind would come and tip up the pram. It was, of course, good weather when I put him out, but the weather could be whimsical."[332] Dagny says she thinks that it is not the usual practice to let children sleep outside in Finnmark. She comments to her mother-in-law about Ole thinking that Oa was much bigger than children of the same age, "but that is not so strange, because they have such barbaric customs about not having their children outside while they are small, which is a great pity when the weather is lovely."[333]

Synnøve, Dagny and Ole's daughter, was quoted in the newspaper, "*Finnmark Dagblad*" in 1983 as saying that not only babies in prams had to be watched carefully, "[…] the grown-ups also had to watch over us when we were playing outside. The storms on Haldde could be unpredictable, one never knew when they would start." Gudrun Dagenborg, the maid on Haldde, supported this, "The children were absolutely not allowed to play outside on their own, because a sudden wind caste, could blow them over, and the ground was stony as well as being steep on nearly every side of the house."[334]

---

331 Memoirs for grandchildren written by Dagny, 1976.
332 Memoirs for grandchildren written by Dagny, 1976.
333 Letter from Dagny to her mother-in-law, 10.11.1917.
334 Nielssen & Petterson 1993:48.

Oa's play area outside the door. Steep sides surround the small plateau. Photo: Ole Andreas Krogness. Archive: National Library of Norway – photo collection.

The children are mainly kept inside, and only go out when the weather allows it. The many days spent inside are at times quite a trial. Both Dagny and Mrs Devik mention in their letters that Oa is an active boy, and he remained the same throughout his childhood, "He is terribly wild and uncontrollable, especially when he has to be inside for weeks on end, poor thing. I wish many a time that I lived nearer to all of you so that he had a bit more company."[335]

So that Oa can get some exercise in the days when the weather is good, they have to make sure that it is safe on that rugged patch of ground. The

[335] Letter from Dagny to her mother-in-law, 18.02.1917.

story about how the boy was tethered fast in the yard outside the house circulates in the family. Oa was bound fast by a rope that was tied to a pole. This restricted the area in which he could play, and safeguarded him from falling down in the loose stones. To make it possible for Oa to have a bit more "free room" where he could play, they fenced in an area with chicken wire. "Here, thank-heavens, everything is well. Good weather and the boy is out in the "chicken run" which we have made for him."[336] Köhler made use of the same method with his children, when he became the Director, after Ole left Haldde.[337] His maid, Gudrun Dagenborg, also tells us that they tethered the small children fast to rope when they wanted to play outside, or built a cage for them.

That a steep-scree plunged straight down outside the main entrance, made it understandable that the parents were worried. Nevertheless, it must have been challenging for the children not to be able to move completely freely, but were restricted, either by the rope or the cage of chicken netting. Maybe it is not surprising that Oa was regarded as restless and wild, when for much of the time he had to be still, and had limited opportunities for physical activity?

Dagny is aware that she is drawn between her role as a mother and her daily work. "I get so little time to tell him fairy tales and care for him, there is always something that has to be done."[338] Oa demands quite a bit of attention from his mother, but it appears that Synnøve is an easier child, when she is well:

> Synnøve walks and does things all day, she is no trouble, and must take after Ole in temperament, as she is so calm. She has had three choking attacks this winter; they are so terrible when she gets them. I will be asking Doctor Gjessing to come up as soon as the weather gets easier. She has this hoarseness all the time, and if she gets anything stuck in her throat a mass of mucous comes into it, and almost takes her breath away.[339]

Synnøve's strangling attacks, as Dagny calls them, are frightening for everyone round her. Especially when the doctor cannot come. After an attack they tried to get Doctor Gjessing to come up, but the weather was too bad. In the years ahead, Dagny has to learn how to tackle the situation of having a little girl who cannot get her breath. At that time there was little else but a steam bath, or a cold-water compress to relieve the attack.

---

[336] Letter from Dagny to her mother-in-law, 18.12.1917.
[337] 1918-1926.
[338] Letter from Dagny to her mother-in-law, 18.02.1917.
[339] Letter from Dagny to her mother-in-law, 18.12.1917.

There was no medicine for asthma attacks then that could help to open the bronchial tubes, and strengthen the mucous membrane.

Members of the household have a coffee break. From the left sitting: Dagny and Ole Andreas Krogness. The other two: the assistant Othar Hoel and his wife. Archive: National Library of Norway – photo collection.

When one lives so closely to each other, the difference in a family's culture is shown clearly. One area we know about, which could provoke discussions was the upbringing of children. Mary Lukkassen told us that she was spanked when she was young if she did not do as she was told. This type of correction was not used on Oa, because his parents were against physical punishment. Dagny herself had grown up with physical punishment, and had as a young child received her share, something she describes in her memoirs:

> My father was extremely kind. However the boys were spanked, as expected, if for example they came late for meals. Our mother used birch sticks and spanked us small girls. That was the normal thing to do to your children at that time. I was apparently quite wild and often got spanked. We lived in Korsvold, the Coach Inn, in Dombås. In the cellar, Guttorm had his bakery under our living room. He heard us screaming when we were spanked. One day I had a new dress, and went down to show

> Guttorm. I sat on a tub with syrup in it, and suddenly I fell into his syrup in that lovely dress, "Now Dagny will have a good beating," Guttorm said, but he came up with me and begged for me!

Parents who chose to physically punish their children at that time had the law on their side. At the end of 19th century the law had strengthened children's rights. Corporal punishment was limited to a child's educational upbringing as stated by the law 20th June 1891,[340] "Parents or legal guardians are given the right to use moderate corporal punishment to further the educational upbringing of children that are under their authority."

Punishment using spanking or boxing ears, which complied with the law concerning educational upbringing, was not regarded in 1915 as ill-treatment, but as, "moderate corporal punishment". Even though there was not total agreement on Haldde, it seems that well-being, and solidarity overshadowed disagreements.

Every opportunity they had was made use of for social activities that lightened up the day. If the weather was so bad that they were unable to go out, they found something that would cheer up their existence. Dagny Devik tells about this:

> There was never a hint of, "just a small family party or such," everything was celebrated. They came down to the party from the Castle, which was the old observatory where they worked. They were all dressed up. "Father, came in large boots, brown trousers, and he had been given a dinner jacket, and I had flowers, artificial ones. And songs were written, and we had such a lot of fun. You have to remember we were young, yes young![341]

Days when there was a reason to celebrate were really celebrated, but there were other days where they found excuses for celebrating and suddenly there was a party. "Today Magda baked a puff-pastry tart, and as we will be having chocolate tomorrow morning for the whole colony, we just have to have a bit of a party."[342]

Even though Dagny and Ole, right from the beginning, have taken every opportunity to turn whatever they could into a party, now their social life is quite different. The house is bigger and Dagny has more people to talk to, and more to visit. Ole's sister Anna has the impression that the social life there was flourishing, and writes the following:

---

[340] 20th June 1891 a new law "*Indskrænking i Anvendelsen af legemlig Revselse*" From 1st paragraph of the law, about limitation in the use of corporal punishment.

[341] Conversation with grandmother, Dagny Devik, 27.08.1985.

[342] Letter from Dagny to her mother-in-law 13.11.1914

> There are more people up there this year than before, a whole colony, so that Christmas will be very festive. Oa has grown so much now, that he is probably looking forward to Christmas, it would be lovely to see him [...], and little sister I hope is fine. You absolutely have enough to look after, Dagny.[343]

Women friends together round the table. Photo: Ole Andreas Krogness. Archive: National Library of Norway - Photo collection.

The families' invited each other formally to mid-morning coffee, either all of them together or individual families. Not long after the Devik's arrive, Dagny describes it like this, "I was invited to coffee by the Devik's on Sunday evening, great fun visiting someone, and I can talk with them about everything, and we also have mutual acquaintances in the south of the country."[344] She does not just drop in here, or is called in because she happens to be passing, but is formally invited to her neighbour for coffee on quite an ordinary Sunday.

When they experience that it is absolutely time to have a party, they do it in style. There are often songs that they have written, and poems or

[343] Letter from Anna to Dagny Krogness 09.12.1915

[344] Letter from Dagny to her mother-in-law, 03.10.1915.

riddles. Dagny tells her mother-in-law about her birthday, which she celebrates despite problems with the electricity:

> Ole and Devik struggled with the electricity, so we were often in the dark and without any heat. On my birthday we had to turn everything off, but finally got a bit of light. I had everyone for the evening meal, and as you know some light was needed. It was an exhausting day even though it is on Haldde. Devik and Tveten have written some verses and some riddles, which were shown as slides, and that was very amusing.[345]

At one of the parties the residents had together, Olaf Devik wrote the following introductory verse for a poem, especially for that evening. In it he shows the most important ingredients for getting together, despite the storm outside.

> This evening it is festive
> Time goes, time goes
> Piano, cello, wine in glass
> No one here suspects
> Outside are huge banks of snow
> The storm rages, but time goes on
> The party's effervescence reaches the roof-tops.

It appears that it is the women who are responsible for the good food and drink, while the men were responsible for the entertainment. Even so, in one of the letters it shows that the men could also make food for parties without help from the women. We can read this in Ole's letter to Dagny, when she and the children have left the mountain for the summer:

> [...] and we continue with the party spirit as well as we can. We invite each other reciprocally to big or small parties. Lukkas began; as I mentioned last time we had a party, with porridge one evening. Then it was my turn to celebrate with sponge cake and orange squash.[346]

They not only arrange parties for themselves on the top of the mountain, but Devik has found an artist, Godal, to paint a picture of the view. Ole continues:

> Then Hoel came and his wife on the 23rd together with Godal, Devik had appointed him to paint the view from the top of the mountain. He had his paint-box, brushes and canvas with him. We made him an easel,

---

[345] Letter from Dagny to her mother-in-law, 26.11.1917

[346] Letter from Ole to Dagny, 31.05.1917.

> Unfortunately the weather was, as usual, not particularly favourable, thick fog most of the time, so I suggested that he could be practical about it and paint just one grey clump of fog on the canvas, as it would be undeniably fairly characteristic.
> Then we quickly found a photograph. He sat in the living room by the window one morning, while I got on with other things in the office, and when I went into him sometime later the top of the mountain was finished. It was quick work and easily done. It was a completely unnecessary long journey just for that painting, even though it was pleasant.

Big events were properly celebrated, as Olaf Devik tells his family in Kristiania in a letter about the christening of both Finn and Synnøve, in the summer of 1916. The christening was celebrated on a summer evening with good friends at Elvebakken,[347] "Doctor Gjessing and his wife, the telegraph manager Feyling and his wife, Captain Carlson and his wife and Godal". The Heitmann family were also supposed to come, but were unable to attend. Ole and Olaf made small rhymes about all the guests, who were then to guess who they were. He wrote about Feyling, who had won the lottery and his wife who played the piano:

> He plays the lottery
> She plays on something with music in it.
>
> To the mothers:
> Both are mothers
> Both are fathers
> Both have the same name,
> Both are beneficial.
>
> To Heitmann they had prepared this:
> He owns a farm and ground and church too
> And deep holes that in the mountain slide down.
> His warehouse contains large quantities
> Of buttonholes, cork bungs, and so much more, in it then.[348]

[347] A village not far from Bossekop. Alta airport is there today.
[348] All the riddles and songs are part of the papers owned by the family etc.

Synnøve Louise and Finn were christened on the same day. There was a big party afterwards. Front row: second from the left Dagny Devik holding Finn, and to the right Dagny Krogness with Synnøve Louise. Archive: National Library of Norway - Photo collection.

Such an evening also had to have its own composed song, something for babies being christened, with rhymes and many verses. Olaf Devik wrote the following song 9th July 1916:

Swedish Music, "*Der gingo tre flickor*" (Three girls they did go …)

At an altitude of 900 metres,
Way up to the everlasting snow
…in the land where the Sami are herding
reindeer that one often does see ...
there is a top, to which one can look up
from Elvebak, the fjord and Bossekop,
and heavy work one feels, when over sticks and stones
one crawls way up, OK, Ok, OK!

And if one has come to Haldde
So one wonders once again:

One meets those who have chosen
To listen to the song of the wild,
… and to follow their men, who flew from the world,
to live where hurricanes speed, speed, speed.
Where Northern Lights and winter's pitch black.
Gives life there, its own special type, type, type.

Greetings, housewives from the mountain!
For you we will sing a verse,
And the songsters will pay dearly
And their voices must face it
For to move up with your husbands to the top
Demands a very special courage, courage, courage.
We thought that today you really deserve
A three times, three times, three hip, hip, hurrahs!

In nineteen hundred and fourteen
A day in September it was
Mrs Krogness cannot have forgotten
And neither Tulleruska's Pa,
As for the first time a newborn lady sang
An aria with her small voice
Melodious it was, inherited from her Pa
For Synnøve Louise has an ear for a tune.

On Haldde the stork is outdated
In the "*kibben*", children are sent up
And next time with good luck
To the Deviks, a small boy was sent.
When he an adult becomes, his mother will surely tell
About the mountain top far in the north
Raised high above the earth, where science dwells
On Haldde Mountaintop whereof there are words, words, words.

Finding a name was a long struggle
And Halldis Aurora was mentioned
If the wives had given in where would it end?
There was also a fuss about Haldor Eolus too
So finally it ended with this, this, this
That both were given names that will not help
As a reminder of their rightful birthplace.

And now they are going to grow up
And soon remember no more
Of all that they as small children did do
for one forgets what one no longer can see.

Upbringing soon there will be, for the sake of love
When early punishment takes hold
Possibly they will be pleased when it is no more,
But discipline must all have, at least once, ha, ha, ha.

Ole took lots of photos and was always looking for a good motive. From the left standing: Dagny Krogness, Dagny Devik and Sanna Pedersen. Archive: National Library of Norway - Photo collection.

There was always a lot of music round Ole Andreas Krogness wherever he might be. Although it was not just Ole that liked music, it was also important for Dagny. It was impossible to form a quartet on Haldde as Ole would have liked, but he was always looking for sheet music and participants. It was not unusual for him to hand out music, which colleagues and friends could read and practice until they met later. "In Hammerfest there will also be a great deal of music, and in the summer the music will possible continue, when two of the musicians come to Alten. They will be practicing, Sinding and Moskowsky, for now."[349]

Dagny felt that she could now contribute to these social occasions. "I have begun to play together with Tveten and Ole, who both play the cello,

[349] Letter from Ole to his mother, 10.11.1918.

and it seems to sound fairly good especially after I have received some corrections and reminders.[350]

Oa also enjoys music, and tries as hard as possible to contribute. As a mother Dagny is also concerned that he should develop his musical ability, and does her best to stimulate him. She continues to write:

> Oa likes the flute, but Ole plays on it most of the time. I think that Oa has more of a musical ear now as he sings the single tones clearly. I have sung Christmas Carols for them every evening for three weeks, and we also sing, Margrethe Munte's songs, which they like very much. […] I thought this would help them have a good ear for music.

In their daily life music is part of the home for Dagny, Ole and the children, and in festivities it is a welcome contribution.

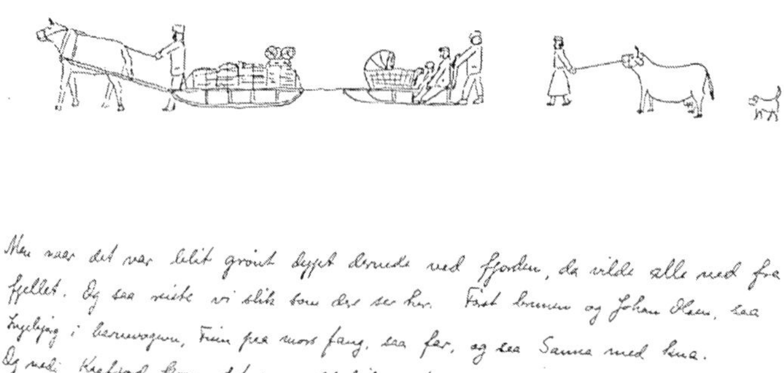

Olaf Devik made a picture book, using his own drawings, for his children about life on Haldde. Drawing: Olaf Devik. Archive: World Heritage Rock Art Centre - Alta Museum

350 Letter from Dagny to her mother-in-law, 01.11.1918.

# CHAPTER EIGHTEEN

# GOAL IN SIGHT

Flag flying. Here there is something to celebrate. Photo: Ole Andreas Krogness. Archive: National Library of Norway – photo collection.

*You are probably surprised at my new handwriting. As you can see we have got a typewriter here, and some practice is needed so that it can be made use of. I dictate and Ole writes.*[351]

About a year after Devik moved to Haldde, they decided to start work on planning a geophysical institute, not just in Bergen, but also in Tromsø.[352] It had not been possible to do anything about this previously, because the government refused to grant money for building projects during the war. Ole had been to Tromsø, and became acquainted with the

[351] Letter from Dagny to her mother-in-law, 10.11.1917.
[352] Devik1971: 76

chemist C. Ulstrup Dahle, who was of the opinion that it should be possible to start a private fund collection towards the establishing of an institute. The government could then take over the running costs. This will have a great significance for the development of weather forecasting services for the region; and at the same time create a department for Northern Lights and terrestrial magnetism. The initiative for this step was that a governmental department had signalled an interest about whether the investigations on Haldde could be of significance for an eventual storm warning system, for the region north of Bodø

When Birkeland had applied to the Ministry of Church and Education for money for the extension to the observatory on Haldde, he had argued that permanent observations on Haldde would be of great help for meteorologists. It would also be of practical significance for fishermen in the northern regions. During his visit to Haldde in 1910 he had observed that some cloud formation had a striking resemblance to the shape of the Northern Lights, and appeared often simultaneously as magnetic storms and earth currents. He believed that he had found an electric connection between the earth's magnetism and meteorological phenomena. This confirmed that the observations of the magnetic variations at higher latitude would have significance in the calculation of a weather prognosis.[353]

It was easiest for Olaf Devik to leave his work on Haldde, and he was given the task of investigating the possibility of getting support for their plan for a geophysical institute in Tromsø, together with a weather forecasting service for Northern Norway. They had to arrange meetings down south, first and foremost with the government. Olaf left Haldde and stayed some days in Tromsø. There he discovered that they were preparing to have a private action to collect funds. In Bergen the fund collection had already resulted in one million Norwegian Kroner, and they estimated that it was certain that they would have a geophysical institute there.

When Olaf arrived for the meeting with governmental representatives to discuss plans for Tromsø with Undersecretary Hougen and Cabinet Minister Løvland, he understood that they were both extremely interested. If the citizens of Tromsø could find a piece of ground and erect a building, the Government would allocate funds for the running costs of a geophysical institute. He was informed that the institute had to establish a cooperative agreement with the Meteorological Institute in Kristiania, concerning instructions, when the weather forecasting service had got started. All they had to do was go ahead with the preparations.

---

[353] Devik1971: 35

Ole and Olaf started an intense campaign to influence opinions and collect funds to build the institute. They wrote articles in the newspapers, they did interviews, and they held open lectures in Tromsø. They enlisted spokesmen for their case in towns in the north, in Tromsø, Bodø, Harstad and Hammerfest. The campaigns' fund-raising committee were citizens of the different towns. In Tromsø the members of the first committee was: Chemist C. Ulstrup, Consul Joh. Rye Holmboe, Chief County Officer Klaus Hoel and Brigadier Johan Nyquist. They collected 46.500 Norwegian Kroner; Tromsø City Council gave the site on which to build the institute.[354]

Dagny's mother-in-law in Kristiania was kept informed. On 23rd February 1917 she writes to Dagny, and tells her that Richard Killengreen from Tromsø, her daughter's father-in-law, has now been in town. Among other things he talked about present day problems, and that everything takes such a long time, and there were delays both with electricity and coal supplies because of the war. He himself has had to delay his journey home. She had visited her daughter, Anna the previous evening and Killengreen was also there:

> I asked him if he believed that the building of an institute would go fairly quickly, and if it would be ready this year. He thought it would be possible, because he had recently heard that 45.000 had been collected, and not all the action funds had come in yet. He had by the way nothing to do with it, as he was away at the time […], but he had told Anna that he had just given 2.500 towards it. I thought that it was interesting to hear that they want to get started as soon as possible, so we just have to hope for the best.

Ole supplements what Killengreen said in a letter to his mother:

> All is well with the institute in Tromsø, and 46.000 kroner has been collected. They are hoping for an additional 15.000 from Tromsø Municipality, the Wine and Liquor Cooperative, and the Private Bank, with 5000 from each of them. Anyway, they will start building this summer. They have thought of constituting a "Scientific association", which should function initially as "owner" of the house. It would mean that they could borrow what was needed if necessary, and not concern themselves with other kinds of negotiations, which could possible delay the work, and may result in a postponement of the whole project for a year.[355]

---

[354] Devik1971: 76-77.
[355] Letter from Ole to his mother, 02.03.1917.

An association for the Geophysical Institute in Tromsø was founded, with Rector L. Stav, as the Chairman. The country's leading geophysicists had come with some expert opinions, and 30th April 1917 the association offered to erect the building and place it at the disposal of the Geophysical Institute. On 19th May a proposition was presented to the Cabinet Minister, which Olaf personally wrote after being requested to do so; sounding like this:

> The Government finds that the submitted plan has been given unanimous approval by experts. It therefore demands a natural continuation of the work that has been begun with the investigation of Northern Norway's geophysical situation, and is in agreement that it is a national responsibility to further this investigation of which its scientific and practical significances are clear.

Procession in the streets of Tromsø. Ole already knew the city well from a number of visits. Maybe he looked forward to being in a lively city centre? Photo: Thoraif Øien. Archive: Perspective Museum.

The government warmly recommended that a geophysical institute should be established in Tromsø. The responsibility for its maintenance would rest with the government.[356]

It appears that Dagny has left and gone to the capital for the summer. She receives a letter from her husband, who is on Haldde.

> My dear wife,
>
> The institute in Tromsø seems to be progressing. The Government has unanimously voted for the proposed bill. It has arrived in Tromsø and appears to be very good, and with copious contents. You should take a walk, preferably to Gröndahl's bookshop down in Toldbod Street, and buy one. It is apparently an excellent lecture to read when travelling. If it is possible could you kindly get thirty copies, and if they are cheap get fifty, and send them to me, that would be wonderful. You need not pay the bill yourself, send it to Obs. You understand, they were intended as propaganda to encourage people to support the fund collection. I believe you can just ask for the Royal Proposition for the Geophysical Institute in Tromsø dated 19th May.[357]

Ole also tells her that the building of the house for the institute is progressing:

> The State architect has now made some sketches partly in collaboration with Devik, when he was there. The house is now quite different, fine and distinguished when such a specialist is used, and now I hope he will soon be finished with it. We hope to get the house completed in October, so I am not despairing at the moment. We must be prepared to help things along if it is to be finalized. Although I think that all will be well.
>
> Now I must close. I have written thirty letters today, mainly bookkeeping work.
>
> Warm greetings to you all
>
> Your affectionate Ole[358]

Everything went well and on 1st June, "*Stortingets budsjettkomitét*", [the Parliamentary Budget Committee], delivered their proposal:

---

356 Devik1971: 77-78.
357 Letter from Ole to Dagny, 31.05.1917.
358 Letter from Ole to Dagny, 31.05.1917.

> The Committee is in agreement that the unusual conditions for meteorological and oceanographic research, as well as the special Arctic geophysics that the region in Northern Norway can offer, should be further exploited, when establishing a geophysical institute in Tromsø.[359]

Ole can relax. He is satisfied with the contribution they have made. Summer is just round the corner, and as usual the colony will leave the mountain. Dagny has already left with the children, but Ole has remained. First of all he has to complete some work. He has decided to stay for the summer in Kristiania. He has quite a bit to write, and it could be an advantage to work with it in the University library, because the library on Haldde is still extremely inadequate. He considers staying at a hotel, because he thinks that, "Mother Guldberg", has more than enough to do. It is quite likely that Dagny and the children are there. But he has not decided when he will leave. He writes to Dagny:

> You ask me about my plans. I had thought that it is probable that I will be coming down south in July. My Donna Olaug will be here, as arranged, until Mid-summer night. By the way, (since we are talking about her), I think I will ask her, when she leaves, if she would like to come with us to Tromsø, unless Sigrid wants to. She is absolutely more clever and easier to cope with than all the others we have had. Anyway, we will cross that bridge when we come to it.[360]

He tells her about a guest who has been on the mountain and has also taken him out skiing as often as possible, "and which has presumably had a welcome influence on my *Corpus Corpulentes*."

He hopes that nothing will prevent his journey but he "would prefer not to stipulate a particular day for when that might be, even though he longs to be with her; but just now she will have to resign herself to his rather unclear plans. He is pleased for them that they are in a southerly region, because, "it is good for you all to warm up your blood, and for the boy, not least to move in the *hautes rollées*, (that is in the more highly educated circles)."[361]

He is extremely well, now that summer is approaching the top of the mountain. This is evident in Ole's letter to his mother later in the summer:

---

[359] Devik1971: 78.
[360] Letter from Ole to Dagny, 31.05.1917.
[361] Letter from Ole to Dagny, 31.05.1917.

Ole celebrates spring by taking a bath. Dagny told her grandchildren that she would not use that "Roman bath". She was afraid that the electrical elements they used to heat the water carried the electric currents through it. Archive: The World Heritage Rock-Art Centre, Alta Museum

> Here summer has also just begun to announce its arrival. Our water cistern is now full of water, which is not so unusual, but at the moment exceptionally welcome. We can also be extravagant with water for once,

> and just be delighted about it. We have carried the bathtub into the kitchen, and can take a bath, 'just like that'.[362]

While the well-being of summer spread on Haldde, a committee proposition was being written to Parliament. This was dealt with on 15th June 1917, and it was carried by a unanimous vote. The first point in the Parliamentary resolution was, "The Norwegian Parliament sanctions that a geophysical institute is to be established in Tromsø, under which is to be the magnetic-meteorological observatory on Haldde.[363]

In this way the activities of the Haldde researchers bore significant fruit. The weather forecasting services for Northern Norway now had institutional roots, with full endorsement from all quarters.

The war ragged in Europe, but those living on Haldde were not especially affected by it. They sat safely on the mountain top and discussed further developments for their scientific work, once the institute was erected. They meant that there were plenty of research tasks to work with in the region. Maybe they could work for the establishment of a university? They had experienced how inspiring it was to be with a group, who worked with new tasks, and they felt that in the future it would be young people, who would see the value of having their work area in the northern region of Norway.

They were also optimistic when funds were being collected, as they saw that the people in the north showed an interest, and a willingness to support the project. Plans for a university were made public by Merchant Hans A. Meyer in Mo i Rana, an acquaintance of theirs, who wrote an article in the Norwegian newspaper, "*Tidens Tegn*", in autumn 1918. In it he argued that Northern Norway should have its own university, and it should be established in the city of Tromsø.[364]

In the letters from 1917, it is clear that the Haldde era is nearing an end. Even so, the researchers continue to work. They have a new assistant. His name is Tveten and he is described as, "a clear acquisition, extremely interested and very clever."[365] Ole is enthusiastic that he has got such a good co-worker, and that it is a complete pleasure to work with him. Tveten and Olaf are already out on a meteorological expedition. Meanwhile Ole writes to his mother:

---

362 Letter from Ole to his mother, 08.06-1817.

363 Devik1971: 77-78.

364 Fulsås 1993:12.

365 Letter from Ole to his mother, 27.11.1917.

> We have now got our wind measure on "*Beskades*" and we are measuring the wind up there through telephone wires, a new method for meteorology [...] They have begun to build the house in Tromsø, and have put the "bottom frame" on, as the foundation wall was already finished in the autumn. They write that they plan to start building now, throughout the winter, despite the snow, of which there is now quite a bit.[366]

The building project takes longer than they had calculated. This gives Dagny cause to worry about where they will live when they leave Haldde, as the institute in Tromsø is not completed. Nevertheless, they find a solution. In one of her letters from Haldde she writes:

> Yes, luckily there has been a decision about where we are to live in the summer. It is rural, and it is just across the sound, there are cows there and chickens so we will get our milk and eggs. Yes, you can just imagine how elegant we will be, as Ole has bought a grand piano in Trondheim, the piano we have on Haldde cost almost the same. We have thought of charging 1000 for the piano and other things we have left behind. [...]
>
> Thank you for Ole's present he is being kept busy writing, and begging for transport for everything, including his wife, maid and children. We have thought about travelling in July or the last days of June, but that will be according to the snow conditions [...] Oh, I dread the thought of moving, my back is so wretched.[367]

In her next letter, she is lying in bed with sciatica, but brings her mother-in-law up to date:

> They have begun on the second floor of the house in Tromsø, but it is the interior that comes last, although there does not seem to have been much painting done either. [...] Ole and Devik have been on a visit to Øksfjord since they came home, the boat goes there now and corresponds with the Coastal Liner; they have so many plans. We are also going to Øksfjord, and from there it is eight hours to Tromsø, that is half of the time we used on the journey from Hammerfest.[368]

The last letter Dagny writes from Haldde is while she is in bed on 4th June 1918.

> It is so disgusting that I cannot work as I would like to, and just as we are in the middle of moving. Ah, yes, things will be all right I am sure. The

---

[366] Letter from Ole to his mother, 17.12.1917.

[367] Letter from Dagny to her mother-in-law, 28.05.1918.

[368] Letter from Dagny to her mother-in-law, 04.06.1918.

children are really good, brown as Negroes and have red cheeks. We still have winter here with -6° C, so it will be lovely to go down from here and have a taste of summer.

It was during the spring that the colony moved to Tromsø. The local boat puffed its way out into the fjord containing the families, and their goods from their mountain home. Their everyday life, research and scientific work now moved on into a new and exciting phase.

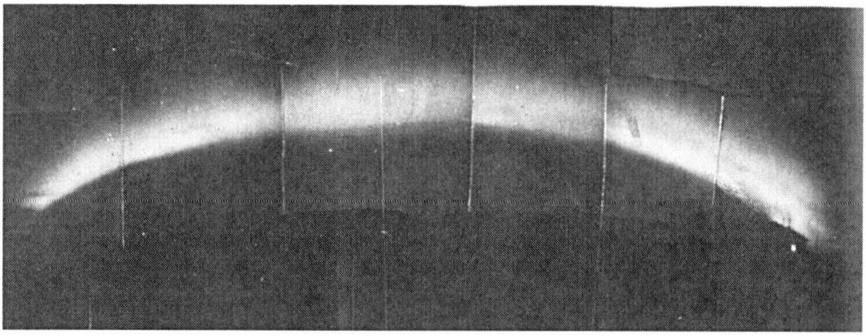

A clear bow of Northern Lights over the top of the mountains. We can just glimpse the house on Haldde on the right. Archive: National Library of Norway – photo collection.

Now the good times on Haldde were over. It had been hard in many ways, but pleasant. The class differences were treated as a matter of course, which was the normal practice then. What was special about their existence under extreme living conditions was that their everyday life had probably had a greater significance, and in all probability a greater respect, as their living conditions absolutely required that all of them contributed. It is possible that the women's presence in the research milieu led to an increased awareness of what was around them, and concerned them, in their daily life. The researchers also had to give priority to practical arrangements, so that everything functioned.

In a later scientific publication, Ole writes in the first chapter that during the spring of 1913, very few parallel photographs were taken that could be measured against each other. After Vegard left Bossekop that year, no Northern Lights were registered from the top of Haldde or Bossekop. Ole explains, in English, that this was due to a great pressure of work:

This was not indeed due to the absence of auroras, in fact they appeared very frequently, but our small scientific staff had so many kinds of

> observations to prepare and to carry out at the observatory itself, that very little time was left for such purposes.[369]

As we know, their situation was so difficult for a period of time that Ole had thought about taking a job in the south of Norway. When the Lukkassen family moved up to Haldde, Ole had an assistant to relieve him of the practical work, but at that time there were already new plans being developed. This involved putting more emphasis on wind and weather forecasting, and in addition studying terrestrial magnetism and the registering of the Northern Lights. Tromsø with its workplace in a geophysical institute had started to blink alluringly in the distance.

---

[369] Vegard & Krogness 1920:5.

# CHAPTER NINETEEN

# HOUSEWIVES AND MAIDS

Christmas on Haldde. Archive: The World Heritage Rock-Art Centre, Alta Museum

*I am so happy that I will soon be rid of the maid. She is so irritating, just sleeps all day.*[370]

There is an on-going theme in Dagny's letters from Haldde. Not about the weather, which one might have expected, from someone surrounded by such hard weather conditions, even though there are many descriptions of storms and hurricanes. It was not about music, even though Ole is very musical and they had transported a number of instruments up to the house, which they often played. It is not even about the children and their births, and it is not about research activities. What is mentioned in almost every

[370] Letter from Dagny to her mother-in-law, 27.12.1918.

letter, are the maids, who are mainly from the nearby villages There is little doubt that a woman's life on the top of a mountain was hard work.

Let us take a closer look at the situation between a housewife and her maids at that time. It was a well-known relationship in the contemporary social system, and had a central role in the forming of everyday life, even on the top of a mountain.

Dagny writes to her mother-in-law, giving us the opportunity of participating in the communication woman to woman, about essential conditions in their daily lives. Within this subject there are many implications concerning culture, and the meeting between different social classes. Everything here happened in the early 20th century, a time of many changes. There was a growing awareness of class differences, and the growth of the labour movement became applicable in several areas of society; not least in relationships between housewives and maids. Much of this is described in academic literature.[371] There is a lot of agreement between the portrayal of everyday life from the beginning of the 20th century, and the contents of the letters we have access to, both with regard to practical work and its organisation. Life on the mountain was not so exceptional that contemporary women's values were not adhered to, as anywhere else in Scandinavia.

As we have heard Dagny and Ole took the maid with them to Haldde. In her first letter to her mother-in-law, Dagny writes:

> The maid is fine, clever and wants things to be clean around her. At the moment she is living upstairs, where there is a large room, almost large enough for two. It is important that she likes it here, because there is a lot to do. We melt snow continually, and as you know that can delay our work considerably. I try to help, but unfortunately my back cannot tolerate very much.[372]

She also tells about what the house looked like when they first arrived, "The house is quite a mess, because there have been two girls who have looked after the workmen and the assistants, and they have helped themselves to towels and tablecloths."

Dagny comes from a wealthy family and was used to having servants whose job it was to help the housewife to get the work done. In her

---

[371] Among other things in Tinne Vammens book "*Rent og Urent*" (1986) from Copenhagen and in Sølvi Sogner's and Kari Telste's book, "*Ut og søkje teneste – historia om tenestejentene*" (2005), Historical account about life for maid servants, describing the Norwegian situation.

[372] Letter from Dagny to her mother-in-law, 02.10.1912.

childhood home they had two maids, as well as a governess, a cook and a nurse, in fact a hierarchy of servants where the "maids" were the lowest on the ladder.

Like most girls at that time, Dagny left and went to the capital, not to find herself a job as a maid, like the majority of village girls, but to work and live with her aunt in a lodging-house for students, called "*Studenterhjemmet*". In Kristiania the influx of village girls was so great that at the end of the 19th century it was called, "*den unge piges by*", (the young girls' city). Dagny works as a receptionist for her aunt. Naturally at that time there were mainly male students lodging there, and the majority of them came from upper class homes where they were used to things being done for them.

When Dagny was about to give birth to her first child, she thought she would need an extra maid, even though she thought it was a disgrace not to be able to manage with just one. She was uncertain about how things would go during the birth, and afterwards. There would, of course, be enough for her to do when the baby arrived, and she would have to look after it, in addition to the usual housework, which was demanding enough as it was. They would also need more water every day.

Ole was kept active shovelling snow into a container to be melted, and Dagny boasted about his hard work. However, nothing is said in the letters about who carries the water into the house. The maid's work is not always specified in the letters. What we do know is that it was usually the women who carried the water. They had a yoke on their shoulders with buckets attached, which were then filled with water. In families that had servants, this was usually a job for the maids. It was not an easy task. Many a woman's back suffered from damage through such wear and tear. One can imagine that in the winter the transport of water was not very pleasant, especially if the buckets splashed ice-cold water, and it ran down their legs. A story from another part of the country tells about a maid, who went into the house with the water-buckets, and her skirt was frozen so stiffly that it could stand on its own.[373]

Already at the beginning of 1913 Dagny writes to her mother-in-law, "We have got a pump in the kitchen, so it is almost as good as living in town. We melt the snow in a tub and pump water up to the kitchen."[374] The problem of carrying water had been solved and Dagny was satisfied.

[373] Sogner & Telste 2015:11.
[374] Letter from Dagny to her mother-in-law, 08.01.1913.

Conversation in the sofa nook. Dagny is on the left. Archive: National Library of Norway - Photo collection.

It was hard for Dagny and the maid to manage all the housework on their own, also before the baby arrived. They needed help with the washing. On 23rd February 1913, she informs her mother-in-law, "I have employed a woman to do the washing up here. The big pile of washing has been in a box for two weeks. Thank you for washing the baby clothes. I have nowhere I can dry anything. That is the worst." She had received new clothes and equipment for the baby, which her thoughtful mother-in-law had washed beforehand to make it easier for Dagny, as she had no attic in which to dry things.

Summer in Elvebakken. Archive: National Library of Norway - Photo collection.

Often it was the girls from the villages who went into town and got a domestic job. They met, what was for them, an unfamiliar town life and had to get accustomed to it. The situation on Haldde was the complete opposite, here it was people from the towns that had moved to a remote area and met unknown cultures. Dagny with her middle class background met the North Norwegian working culture through her maids. This is made clear in her letters. Class differences become apparent in Dagny's description of how Aminda, the maid, reacts to her new life:

> The silver bowls and centrepiece glitters, and Aminda is excited about everything being in its right place. She forgets the washing up and everything else when she can put things that are new, into some order. The kitchen is extremely elegant; she has hung up all the tin forms starting with the small cake tins, up to the large covers, which she begged for. It is really amusing that she wants everything to look nice, but I think it is mainly vanity to show those from the village how nice everything is for her here.[375]

Dagny thinks the girls in the north of the country are so very different from those in the south. At one point she has a maid with almost too good a humour, and too little seriousness. In addition, she is "very scatter-brained. She does not think, so that she often has to run back and fetch things she forgets." She is also talkative, and preoccupied with showing how much she likes it here, because down in the village people talk about us. Dagny writes to her mother-in-law:

> They do not have much to talk about, except us; what we eat, if the food is good and whether we have nice things in the house. Luckily the maid is overwhelmed by so much that I hope they will not be ashamed of us.[376]

In non-fiction literature at the end of the 19th century there were discussions concerning changing the gentry's attitudes towards servants. The cookery book author, Marie Kübler,[377] talks about "the servants' increasing emancipation". She was among those who pointed out that maidservants had a numbers of deficiencies, which she believed to be the cause of the crisis between the gentry and servants. Serious accusations were pointed at young girls who earned their livelihood with rich families.

> Maids were accused of being insubordinate, obstinate, gossips and thieves, promiscuous and useless, unstable at work and with an inclination to dress up and behave as if they were real "Ladies" ... To transform girls and point them in other directions was therefore a main concern, and a primary task for housewives to be responsible for, both inside and outside the home.[378]

Whether Dagny undertook such a "transformation task", with her maids is not directly evident in the letters. Perhaps it was unnecessary to

---

[375] Letter from Dagny to her mother-in-law, 31.10.1912.
[376] Letter from Dagny to her mother-in-law, 19.10.1912.
[377] Kübler 1886, ref. in Vammen 1986:102.
[378] Vammen 1986:102.

do anything about this with the shy and bashful young girls from a local society in North Norway, who probably first and foremost, admired the town people they worked for.

Oa sits contently beside the water buckets. Archive: National Library of Norway – photo collection.

In our material we have an interview with a maid who worked on Haldde after Dagny and her family had left. As a young girl, Gudrun Dagenborg, from Kåfjord was appointed by Ole to work for Köhler, who took over his position as Director on Haldde. The interview took place in the nursing home in Kåfjord. Gudrun told about how ashamed she had been of her lack of knowledge about the world. She barely knew that the world was round, she said. It was also with a great reluctance that she talked about poverty in the village at that time. Even though things had gone well for her, Gudrun was completely aware of the class division between the household and servants there.

It is possible that it was not easy to get village girls to stay on Haldde under the difficult working conditions that life on a mountain offered. Dagny had to handle the girls extremely carefully. Even so a position in a house of a family that was quite well off, at the beginning of the 20th century, was often a first step for young girls in preparation for their own adult life. For some girls it was an opportunity to earn money so that they could take an education later.

The North Norwegian girls that applied for a domestic post, often came from poor circumstances, and a region where the health system was not good. For example it was a fact that one could not rely on dental help when needed. There was a great lack of dentists. A daily brushing of teeth was unlikely to have been encouraged, or practised before public dentistry started up around 1910. Northern Norway was a long way behind such a development as were outlying villages in other parts of the country.[379] False teeth were used from a relatively young age. Dagny told her mother-in-law that the laughing maid on Haldde had such bad teeth that all of them had to be pulled out. That was the end of her being talkative for a while, "But just wait until she gets her false teeth…!"[380]

The maid Dagny has during the first part of her stay on Haldde, knew nothing about cooking, so Dagny had to teach her. "She was good at helping, but was confused by strangers", Dagny writes. They worked together in the kitchen, and Dagny gave her instructions about how things should be done, "It is so strange with such girls because if they get too much help they think they will always get it."[381] As Dagny was thinking about going to Bossekop for the birth, she puts her maid in charge of cooking the meals and reports to her mother-in-law, "She runs at a gallop so I am sure she will be good in time."[382]

It was in no way easy for Dagny to manage the housework on her own. She had tried:

> I was alone from Saturday until today, as the maid was free, and as it is such a long way for her to go it is not just half a day, as it is in Kristiania. I

---

[379] Bore 2007:206.
[380] Letter from Dagny to her mother-in-law, 15.12.1912.
[381] Letter from Dagny to her mother-in-law, 31.10.1912.
[382] Letter from Dagny to her mother-in-law, 23.01.1913.

> thought she was not coming up today, but then she arrived accompanied by two workmen, who are still working here.[383]

The distance to the village made it difficult for both maids and housewives on Haldde when the girls had a free day. Much of their free time went with the journey down to the village and back up again. There is nothing in the letters about how this was actually solved, but most probably they had a flexible arrangement so that none of them would be dissatisfied.

They worked hard on Haldde, and to utilize the days as much as possible, they got up early. The energetic Dagny thought it was fine, "I am so pleased because the house changes character when everyone gets up, and the maid manages to do a lot more, although it seems terrible having to get up so early, but on the other hand we also go to bed earlier."[384]

Dagny's mother-in-law had clearly hinted that they needed more servants in their house on Haldde, because Dagny writes:

> You talk about a housekeeper, but we do not have a room for another person. She would have to sleep in the living room. We had decided that the maid's room was for guests. To have the maid sleep in the kitchen is not very pleasant, and there is only enough room to keep things in order there, as it is very small. The room the maid has now is much too big so two beds could be put in there. I had thought that in the summer we could perhaps make it into two rooms, because it looks as if a lot of people will be visiting us.[385]

The close working relationship between a housewife and her maid, when they often lived in close quarters, did not, nevertheless, mean that they became equal partners. The housewife kept her position of power and authority, and the maids had to adapt to rules determined by others. The maidservants had very little influence on their work situation. The following quotation from Hanna Winsnes tells us something about the kind of instructions and control that was within this power relationship:

> The mistress of the house should keep a record of her equipment, and it is quite useful to take a day once a week, or at least every month, to see if the stoneware and other things that the maid is responsible for are in place. The time you spend doing this is well worthwhile, the maid will be all the more careful handling thing, when she knows that a day of reckoning will

---

383 Letter from Dagny to her mother-in-law, 10.11.1012.

384 Letter from Dagny to her mother-in-law, 29.11. 1912.

385 Letter from Dagny to her mother-in-law, 23.01. 1913.

> come quite often. It is not my intention to make the maids pay for any breakage they may be responsible for, as they do not have the means to compensate for any damage they may have done. A possible consequence of this may result in it being difficult to get maids. There are ways to encourage the maids that are careful and behave properly, for example an extra day off, which is not offered to those who neglect their work. It will be an encouraging incentive for a maid to see that her competence is appreciated.[386]

In our source material we have examples that show that the relationship between housewife and maidservant could be on very familiar terms. For example, Dagny writes to her mother-in-law that the maid has asked her to greet her and say thank-you for the dress material, which Dagny's mother-in-law sent her as a present. She has already made use of it because Dagny writes that, "the dress fits her so well."[387]

The typical division of work in a household, at that time, was that the domestic help was responsible for the heavy work: cleaning, washing clothes, cooking, washing up, ironing and looking after the children, when their mother was otherwise occupied. The housewife decided dinner and kept an eye on it being cooked. She looked after the clothes and linen, polished the silver, bought and ordered necessary wares, looked after the children, and organized social occasions, and perhaps participated in committee work.[388] There is no reason not to believe that this division of work on Haldde kept to these general rules, with modifications that may have arisen due to where the house was situated, the interior layout, and the family's' special situation.

After only a few months Dagny began to long for someone to talk to. Therefore, if she has to have another maid, she would prefer to have an adult maid, so that she has someone to converse with, "It would be so much nicer for me rather than just having a young girl to talk to," she complains.[389]

---

386 Winsnes 1845.
387 Letter from Dagny to her mother-in-law, 10.08.1915.
388 Sogner & Telste 2005:95.
389 Letter from Dagny to her mother-in-law, 08.01.1913.

Quite a few young girls preferred to work at the fisheries. Photo: Olmar Egenæs. Archive: The World Heritage Rock-Art Centre, Alta Museum

Tinne Vammen, a Danish historian, points out how the gentry's children, and youth, early in their lives and living with servants learnt, "that there were people born to command and others born to obey," and that there was a difference between, a "lady" and a "maid".[390] One could have a conversation with a lady, but from a maid one expected nothing else except a well carried out job. That is probably something that Dagny had learnt from her childhood home.

Her plan of appointing an extra maid, that first spring, could, she realised, be difficult to accomplish. "All the girls are employed by the fisheries when a maid is needed, and it is impossible to get any."[391] She had found out that in April they began to dry the cod in the fishing villages along the coast. This created work for young people, and made it almost impossible to find a maid. While earlier a position in a house was the only possible job for young girls in the area, now the fishing industry enticed them with good wages and seasonal work. This was such an attractive offer that the majority preferred to accept this, rather than the alternative of being in service. It was a period when housewives often had difficulty in finding people to work for them.

[390] Vammen 1986:255.
[391] Letter from Dagny to her mother-in-law, 23.01.1913.

New work possibilities had created a shortage of maidservants in the market both in Norway and Denmark. Already in the early 20$^{th}$ century there were relatively few girls in service compared with the numbers a hundred years previously. While nearly half of all 19 years old girls in Norway had been in service in 1801, in 1901 there was hardly a third. The concept, "maid shortage" captures what housewives experienced in the 20$^{th}$ century.[392] There was no longer a flux of maidservants in middle-class families, and housewives suffered greatly because they were unable to manage all the housework alone.

The solution to the "maid shortage" on Haldde, when the maidservant gives up her job in the spring of 1913, was that Dagny's sister Arnhild was summoned to help, which was much appreciated by Dagny. They thought it was sad when Arnhild returned home, even though they tried to persuade her to stay. Luckily after a short time they appointed a new maid:

> Have, by the way, now got a clever and nice girl who can do this and that, and am really pleased that I gave notice to the girl I had last winter. This one has studied at the domestic science school for half a year and has been out in service before. She is of better parents, so she has discretion, and that is so lovely.[393]

The first Domestic Science school in Norway already started in 1865. Housekeeping had become a subject for study. School kitchens were started in 1890. The Education Act opened up for teaching of Housekeeping from 1908, but the education of teachers in Housekeeping had already begun in the 1890's. In 1909, the Norwegian State College for Domestic Science Teachers in Housekeeping was established in Stabekk,[394] and from 1918 the Norwegian State financed running costs. That Housekeeping was given status as a study subject led to the establishment of a trade union, which had a great influence on the situation between housewives and their maids. Both groups had more authority in their role as a woman. It is possible that Dagny appreciated this.

Finnmark's Domestic Science School for Housewives in Kåfjord was established in 1911, the year before Dagny arrived. In the first year there were twelve students, and the course was held about a kilometre from the centre of Alta. When the Copper Works were closed down, and the Director left Kåfjord, the school was moved to the Director's house and

---

[392] Sogner & Telste 2005:777.
[393] Letter from Dagny to her mother-in-law, 23.10.1913.
[394] Sogner & Telste 2005:85.

student numbers increased to eighteen. The Domestic Science School became a cultural meeting place in the village. The clergyman was invited there once a month when he held services. Village folk were also invited to the school for food, drink and entertainment. At the school both Housework and Dairy Work was taught, and for Cookery they had to provide their own raw ingredients. Musical evenings were also arranged there.[395]

The Domestic Science School meant that the daughters from smallholdings and coastal fishing families could have the possibility of being educated, and in turn have a stronger position on the work-market. This also made it easier to get a domestic position in towns, which many preferred. It was not unusual for girls from the villages in Northern Norway to take domestic jobs to earn money for further education in the south of the country. For example, Petronelle Haakegaard, from Kåfjord did this. She worked on Haldde after her confirmation and was later trained as a nurse, most likely inspired by the stories of Dagny's relative, Cathinka Guldberg. Petronelle, the nursing sister, who started her working career on Haldde, has a road in Kongsvinger named after her, "*Søster Nellas vei*".[396]

How aware Dagny was of the situation between housewives and maidservants is difficult to say. Perhaps for her it was just a matter of course in their daily life, which was not questioned. It is unlikely that Dagny knew that the Domestic Science School for housewives was a result of housewives not wanting to lose their position, when maidservants began to organize themselves in a union, at the end of the 19th century.

Originally, "*De Kvinnelige Tjeneres Forening*", the Women-Servants' Association, was founded in Kristiania in 1886, but it had only a short lifespan and there were no traces of it after it closed down. Oscar Nissen, the foreman of Kristiania's Labour Party, went in for the position of servant's cause, with heart and soul. Women working in the matchstick industry went on strike in 1889, and he showed how the phosphorus burnt the girl's faces, and encouraged members to support these girls. Maybe it was such an important event that Dagny, who had grown up in a family of strong women, became acquainted with it when she was an adult?

The young Fernanda Holst, was the leader for the strike workforce. Later her surname changed to Nissen, as she married the much older Oscar, after they had both divorced their respective partners. Fernanda Nissen became an active advocate for the improvement of working women's circumstances. Even though the match factory strike by female

---

395 Interview 17th July 2008 with Rannveig Pedersen, Kåfjord.

396 Interview with Rannveig Pedersen, Kåfjord.

workers stopped without any significant demands being met, it did result in the establishment of a trade union. The match factory strike for workers was a milestone, which also had repercussions for the feminist movement of that time. It was an appeal to other women to become trade unionists.[397]

In 1890, Oscar Nissen again called in a mass meeting for the Labour Party, to discuss the question of maidservants. He was given great support to form a union. When he encouraged those that were present to show their support for a union by standing up, everyone did so in unison. The union was called, "*Tjenestepigenes Særlag*", a special association for domestic servants, and its aim was to raise the social and economical position of domestic servants through regulated work, free time, a proper wage and one's own room. This attempt unfortunately failed, mainly because the maidservants, who were members of the committee, chose to leave the country and emigrate to America in an attempt to obtain better lives there. In 1897 another attempt was made to try and form a union for domestic servants, and a Domestic Servants' Union was established. This time, housewives also took action. That same year "*Hjemmenes Vel*" the first housewives' union, for the welfare of housewives, was established as a counterbalance to the Domestic Servants' Union, and to defend the position of housewives.[398]

Politically the domestic servants, and quite often the housewives, were political opposites. While housewives went to the right, the domestic servants were closer to the Labour movement with regard to their disputes. In Denmark domestic servants that were organized, supported the red flag more and more.[399] Also in Norway, it was through the labour movement that domestic servants were able to uphold their rights with regard to working hours and wage agreements. As late as 1911 the liberal newspapers reported how provoked housewives were, because domestic servants had dared to form a trade union. That servants wanted to stop work at nine o'clock every evening, was regarded as being an impossible demand.[400]

Dagny, as an enlightened woman from the capital city, grown up and married into a scientific milieu, cannot have been completely unaware about the organizing of women's trade unions; especially as she as a young housewife has maidservants herself as work partners. Despite her not having any further education, like her brothers, there is little that indicates that this is a problem for her. In her letters and memoirs we

[397] Ertesvåg 1979:475.
[398] Sogner & Telste 2005:83-84.
[399] Vammen 1986:14, Hølaas 1945:851.
[400] Hølaas 1945:85.

cannot find any documentation that she involved herself in women's politics.

Even on Haldde the residents are made aware that domestic servants have begun to be rebellious and demand their rights. In the autumn of 1916 Dagny sacks one of the girls, because of a lot of trouble concerning wages and working conditions. This is her comment to her mother-in-law:

> Both Ole and I thought, that if she did not care more for us, but was only there for the sake of the money, she might as well just leave. She has 25 and wants 30 in these times. There are no girls in Alta that have less to do for such a good wage. But we have been too nice, that is the truth of it. And now I am so worn out with the children and nervous, that I am completely unable to do any more.[401]

That year the liberal press was concerned about domestic servants' demand to raise wages up to 40 kroner a month. This made it practically impossible to procure maidservants. "The girls prefer other work with more spare time and higher wages, and have not calculated how much board and lodgings is worth." Such comments could be read in newspaper columns.[402]

The new girl who went to Haldde was only 15 years old, and she was not used to housework. Even so Dagny writes that she is tolerably good, except that her knee hurts her after the journey up to Haldde, and so she has been unable to work for a while.

One can wonder whether Dagny exaggerates the difficulty of getting help in the house. To have a position with researchers on the top of a mountain in Finnmark can for some have appeared to be very attractive and exciting. On one occasion Dagny writes about this, "I have been asked if I would like a young woman in the house, who can cook, and not to think about wages, as she would like to go to Finnmark. She is from Bodø and is the daughter of a fisheries inspector."[403]

[401] Letter from Dagny to her mother-in-law, 08.09.1916.

[402] Hølaas 1945, about the year 1916.

[403] Letter from Dagny to her mother-in-law, 27.05.1913.

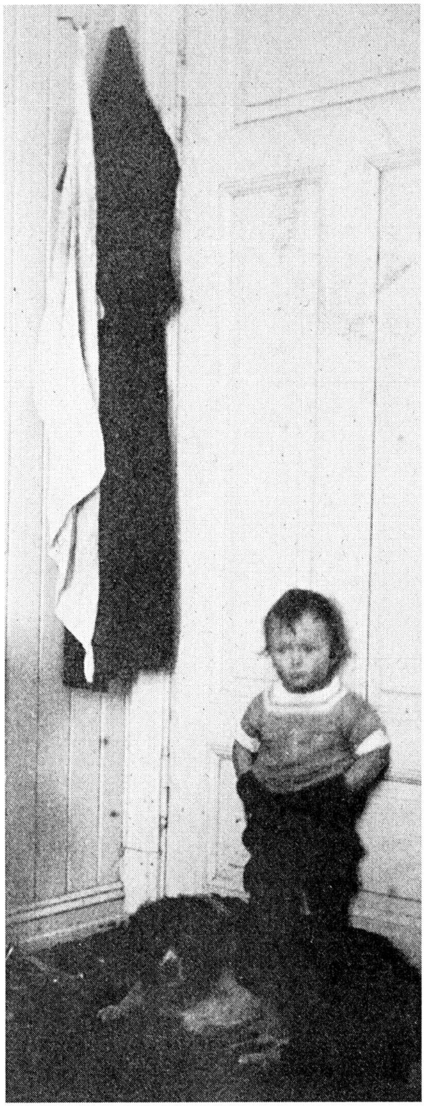

The children were aware when a maid they liked disappeared from their home, and then it was nice to have a dog as a friend. Archive: National Library of Norway – photo collection.

It was an advantage to have maidservants, who had an education and were good mannered. There were also other qualities that were important. They had to be quick, hard working and industrious, look decent and carry out duties conscientiously. They should be orderly and economical. Preferably they should be able to think quickly. Dagny writes this about one of her maidservants, “She has to be asked repeatedly about the same thing”.

It was not always easy for a maidservant to know what was expected of her. There was an instruction book with good advice both to maidservants and housewives. Among the many guidelines we find the following:

> Do not bang doors, speak quietly, and do not sing. Only in absent-minded moments a girl may hum and sing, but it is both thoughtless and tactless. One would prefer to see a cheerful smiling face, but even later in the day one cannot at any time be disturbed by singing and loud noises from the kitchen and halls. The gentry may have work to do at home, which requires no disturbance, and one is also in a position to demand that the housework is done quietly.[404]

When Dagny was in the south of Norway one summer she received an upsetting letter from Ole, who was still on Haldde. He had received an angry telephone call from a furious mother in connection with her daughter’s workload as a maidservant for Dagny.

> She began to complain rather dreadfully. She took it that her daughter was to be a nursemaid and nothing else. She had received a letter from her in which she had reported that she had to work hard both with this and that, and that she was unable to manage to buy necessary things on the wage she had, etc.[405]

The angry mother wanted Ole to ring and say that her daughter should only work as a nanny for the children because she was so young. Otherwise she could be sent home.

Ole continues to write, “You understand I was rather surprised. Took it that it must be due to some kind of nervousness, and as you know people up here generally have a tendency to use strong language.” The sober-minded Ole tried to calm down the situation.

[404] Høegh Schmidt: 1912:13.
[405] Letter from Ole to Dagny, 31.10.1917.

> You must understand she should not exhaust herself too much, because as you know it can be dangerous for girls of that age. You need to talk to her and find out what is bothering her, if there is anything, and so try and put it to rights. Try and encourage her, and make her feel happy and content, that is the main thing. Increase her wage if that will help, a bit more or less during these months is of little consequence. I will be getting 500 kroner rise next year. (ca. 14.984 kroner or £ 1446 in 2016)

After Midsummer's Night the same year Ole left Haldde to travel abroad, and did not return until November, and by then Dagny had employed a new maidservant:

> Not less than five applicants applied, so I had enough to choose from. I made a lucky choice, and found one from nearby the fjord, with whom I am very pleased, at the moment. Devik's maid helped me a bit during the first days here, so all was well.[406]

The two Dagnys had a lot in common. Archive: National Library of Norway – photo collection.

[406] Letter from Dagny to her mother-in-law. 10.11.1917.

Devik's maid, Sanna Iversen, was from outer Kåfjord, and was in service to the Devik family from when she was 17 until she got married to a young man from Kåfjord when she was 21 years old. Sanna's work as a maidservant in a house was, like many other girls, an education in middle-class values and a preparation for their own lives as housewives. If they married well, their economy would probably be enough to appoint their own maidservant.

When we interviewed Sanna's daughter in 2008, she told us that her father often sprang up and down to Haldde to visit his beloved. Sanna used to say to her daughters, "You have considerably more freedom than we had."[407] This remark can indicate that Mrs Devik kept the young couple under close surveillance, so that the moral codes of that time were kept. The boy friend had to spring down to the village again after the visit. It was the housewife's responsibility that nothing improper occurred. A pregnant maid was not wanted for housework, and also the effort of finding another helper was always worrying.

Dagny continued to be pleased with her maid:

> My new girl is nice and clever, so things are going well so far. Today she has cleaned the living room, not the walls and ceiling, because I have to do spring cleaning later when we leave here. The kitchen has to be cleaned thoroughly due to the damage done by the carbon monoxide.[408]

Dagny is worried about how they will manage to do all the cleaning when they leave Haldde. One problem is her back and another the scarceness of water so late in the year. Once again a maidservant problem is mentioned. She has a, "kind girl who is economical, which is a good help in these times. The problem is that she is going home before we leave so I will have to find someone else."[409] The maid had to go home to get a suitcase for the journey, "Ah, yes", Dagny writes, "there are many problems!"

The letters show us how much people from the middle-class, either living on the top of a mountain, or elsewhere, relied on their maidservants. A reliance that became more necessary for each child that was born.

---

[407] Conversation with Åse Synnøve Opgård, 15.07.2008.
[408] Letter from Dagny to her mother-in-law, 26.11.1917.
[409] Letter from Dagny to her mother-in-law, 04.06.1918.

Magda Lukkassen, the maid on Haldde, later became the Manager at a nursing and old people's home in Kåfjord. Archive: Private, Anne Marie Sørensen.

The women's participation in the colony on Haldde, as housewives and maidservants, in a somewhat special work place, made it possible for the research scientists to carry out their work. Dagny accepted living on the top of a mountain, looking after her children, and doing her housekeeping, while Ole went on his journeys in connection with his scientific work, but something she completely relied on was good help in the house. It was also an important matter for Ole.

One can maintain, that much of the quality in every household that had maidservants, relied on the "girls"; exactly as stated in the instruction book for servants:

> The maidservants play a large role in the comfort or the lack of it, in the home. When a home has become what it should be, a cosy meeting place for the whole family, it is the maidservants who should receive a great deal of honour for that. No one should forget that, and remember to acknowledge and give thanks to those who have earned it.[410]

We have no idea whether Dagny would agree with this, or that she would give others the honour of making their home on Haldde a cosy home. One thing is quite certain, that without the maids their way of life, linked to scientific development, would not have been possible. Therefore one thing is certain, the maids represented an important part of everyday life, and gave support to the progress of the scientific work carried out on Haldde.

[410] Høegh Schmidt 1912:38.

# CHAPTER TWENTY

# NO RETURN

Much was uncertain in those times. Archive: National Library of Norway – photo collection.

*You can just imagine that I speculated on whether we should go south or not, because everything is so uncertain.*[411]

Dagny must have experienced worries about the development of the war, but the uncertainty of the world situation did not hinder progress in the research colony on Haldde. On the other hand, the outbreak of the war had extensive consequences for Birkeland. He did not come to Haldde in

[411] Letter from Dagny to her mother-in-law, 18.03.1917.

the summer of 1913, as those living there had expected. Instead he moved to Egypt when his mother died and his marriage was dissolved. The conflict with his friend Størmer and working partner Eyde was also something that Birkeland left behind him when he went back to Helouan, where he had been earlier with Krogness.

His own health was also a factor in his decision to leave the Norwegian cold climate. He now planned to continue his study of Zodiacal lights. He was enthusiastic and full of ideas, as usual, even though the war broke out, and gradually made existence difficult. Among other things he exchanged letters with Miss Ellen Gleditsch,[412] about the treatment of patients with tuberculosis and cancer, with the help of sheets and cloths impregnated by sunrays. He had got this idea from seeing how patients in Egypt were treated with the sun's rays. Wearing few clothes and unconstrained, they turned the sick part towards the sun. He maintained that thousands were healed by the sunshine in Egypt. In a number of letters to Miss Gleditch he suggests that she should send in a patent application for both of them, and start to work with this project. They will earn some money, and perhaps even be given a bit of honour for it, he believed.[413]

In regard to the disagreement with Eyde he writes in a letter to Olaf Devik, dated Helouan 13th June 1915:

> [...] By the way, who has written the biography in the magazine, "*Opfindernes liv*"? It is well meant and written with good intentions, and there is much that is correct in it, but I could make some essential remarks. It is the first time I have seen any mention of the coincidence surrounding the first meeting between Eyde and myself mentioned in Eyde's biography. That account does not come from me, and is not approved of by me. I shall limit myself to just saying that. Essentially the main point about the invention is the tireless experimental work, in which many partook, and which I led through many years. That work constitutes the foundation upon which the Norwegian Saltpetre Industry was founded, and built upon, and the groundwork was so solid that the industry stands firmly, despite a number of extravagances.[414]

---

[412] Ellen Gleditch (1879-1968) a Norwegian chemist with radioactivity and radiochemistry as her field of research. She worked among other things for Marie Curie and at Yale University in USA. Became Doctor of Science in 1914. See Kronen 2009.

[413] Letter from Birkeland to Miss Gleditch, 08.11.1913, and 05.03.1915. Archive: National Library, letter collection.

[414] Devik 1971:73.

Birkeland in front of his house in Egypt. Photo: Ole Andreas Krogness. Archive: National Library of Norway – photo collection.

It is well documented that Birkeland wants to come home to Norway after he has been in Helouan for some years. Isolation began to bother him, and he longs to return to his friends and colleagues, and his good laboratory at the University in Kristiania. He also wanted to participate in the plans that were now under development, by the Haldde researchers. But the ravages of war made any connection difficult in Europe. A journey through the East and Russia was regarded as being less perilous.[415] Birkeland writes to Sæland, "Had thought of coming home last summer, but the war is so troublesome, and I have a wonderful property down here, much more beautiful than the one I had in Kristiania." He gives Ellen

[415] Circumstances surrounding Birkeland's return journey and his stay in Japan. See Egeland &Burke 2005.

Gleditsch some good advice, "Be careful not to stay too long abroad, one becomes so damned homeless, and Norway is the most beautiful country in the world, if one has a good enough health to be there during the winter."[416]

The 8th June 1917 Ole writes to his mother from Haldde that Birkeland, "has apparently reported his arrival from Singapore, but with the situation in Russia as it is now, it is doubtful that he will manage to slip through." A telegram from Birkeland in Singapore had come for Krogness 3rd April 1917. It stated, "Arrival in Tromsø around the first of June. Prepare the plans for the new institute in Tromsø"[417] Ole now knew that Birkeland would participate in the planning of the Geophysical Institute in Tromsø, which he and others had been working on for a long time. So that when Birkeland did not arrive, Ole took it that he was having problems with the journey through Finnmark's neighbouring country to the east, Russia.

We now know that Birkeland died 15.06.1917 in Tokyo,[418] just a few days after Ole wrote the letter. In the middle of the happy events, which were a continuation of Kristian Birkeland's daring establishing of a Northern Lights Observatory on the top of a mountain in Finnmark, his life ended in a hotel room in Tokyo. He was on his way home, with Tromsø as his goal. He had little opportunity to enjoy the results of his work in Northern Norway

According to Birkeland's philosophy, a person's life was a bagatelle in an episode among other episodes in the big picture. A short while before he travelled to Egypt, he had presented such thoughts during a lecture for the Scientific Academy called, "About the world's origin".[419] Here is some of what he said:

> It cannot be contradicted that people, in the last two hundred years, have developed culturally and scientifically much more than in the great many thousands of years that people have lived here before. How long will it last this time? The geological history teaches us that life on earth is a short episode. Henri Pioncaré has said that, "our thoughts are like lightning in a dark night", with respect to the darkness that was over the world, and the darkness that is coming, when the result of all human work will again be lost. But it seems, after what is developed above, that new worlds come into being in outer space more frequently than creatures are born on earth. Most probably each world has its "lightning in the night", with reasonable

[416] Letter from Birkeland to Ellen Gleditch 12.11.1914. Archive: National Library of Norway, letter collection.

[417] Egeland 1994:49.

[418] Egeland 1994:6.

[419] Lecture 31.10.1913 ref. Devik 1971:45-46.

beings who fight, think and discover, as once again they completely disappear and worlds die more frequently than living beings on earth, or more correctly put, in numbers that exceed every boundary.[420]

Birkeland with some small pets. Archive: National Library of Norway – Handwriting collection.

Possibly the war was one of many important causes for Birkeland's death. His close friend and assistant, Karl Devik had been called home to do his military service. Birkeland became even lonelier in Egypt. His health got worse. He was ailing. It is possible that the problems with the testing of the Birkeland-Eyde Furnace had given him a blow.[421]

He left very few signs of his everyday life behind him. An open question is whether the "glue" required for one's existence, was something that the genius Kristian Birkeland lacked. One remnant of his daily life from his stay in Egypt is a photograph of himself sitting outside with a monkey, a large tortoise and a newly born piglet on his lap.[422] Another remnant of his life points towards more domestic occupations in combination with his interest in inventions. He has sent a small wooden box with Turkish caviar to his friend Professor Amund Helland. He

---

[420] Ole Krogness discussed a similar theme in an article in "*Universets utvikling*" (1932).

[421] Egeland 1994:45.

[422] National Library of Norway, Handwriting collection.

wonders if cod roe was treated in the same way, that it could perhaps be both just as good and valuable. “It is pressed and dried and is served like cheese,” he tells Helland.[423] Later that year he sends the recipe for these delicatessens to his good friend.

It is 1916, a year that seems to be inspiring, positive and problematic for Birkeland. He makes a remark to Helland, which is just a hint of an emotionally exhilarating experience.

> I have experienced my life’s greatest adventure here, but the adventure’s features and contours are for me a hundred times more indefinable than the Zodiacal lights are. What I have experienced is however incredible.
> With kind regards. Yours Kr. Birkeland.[424]

Boat trip in Egypt. Archive: National Library of Norway - Photo collection.

It is known that Birkeland had a woman friend in Egypt, who meant a lot to him, the Greek pianist Helle Spandonides.[425] When the war broke out she travelled back to her homeland. The war changed Birkeland’s

[423] Letter from Birkeland to Amund Helland, 27.10.1916. National Library of Norway, letter collection.
[424] Letter from Birkeland to Amund Helland, 12.11.1916. National Library of Norway, Letter collection.
[425] Egeland & Burke 2005, Jago 2001.

situation in Egypt. He experienced that, "Everyone in the whole world is mad, but I have absolutely no intention of being killed for that reason."[426]

He tells Ellen Gleditch in a letter about an on-going conflict, "I am at war with the Mayor in Helouan".[427] It is about a settlement in connection with a tenancy agreement. The Mayor had helped Birkeland find an office and the contract was to be confirmed publicly. The Mayor was not pleased with the payment he received for his service, he demanded more and was given it. Nevertheless, the dispute got out of hand and developed even further, Birkeland explains in his letter:

> [...] suddenly one beautiful day he goes to the police station dragging me with him. I was a German spy or something like that. On the way he asked me, using an interpreter, if I would now pay him the pound. I told the interpreter that he was not going to get a pound, but I would see him hang as soon as I had a chance to do it.

Tempers calmed down and in the next letter to Miss Gleditch, Birkeland writes, "I have settled the dispute with the Mayor of Helouan; do not talk to anyone about it. Apparently, he is not so ill-natured anyway."[428]

As the war was the reason why his research assistant Karl Devik had to go home, all the work was left for Birkeland to do alone. In his letters he complains about everything he has to do and about his health. The heat in Helouan is unbearable in the summer. "One develops stones in the kidneys and liver and the heart at times, and rheumatism everywhere, as one is unable to move in such heat." Then he suddenly has a wonderful idea, which he tells Ellen Gleditch about. He is going to build a museum.

> (...) a small museum for "natural philosophy" which will be unique in the whole world. It will be a reminder of discoveries about the origin of the earth's magnetism, the nature of sunspots and of the formation process of planets. Now just listen to this! On the top of a small mountain rock I will build a dome using the best kind of granite and cement it in place. The walls will be a metre thick and the floor will be the actual mountain itself. On the top of the dome, 14 metres in diameter will be a gilded copper sphere, 3 metres in diameter surrounded by a thin metal disc 7.5 metres in

---

[426] Letter from Birkeland to Ellen Gleditch, 08.11.1914, National Library of Norway, letter collection.

[427] Letter from Birkeland to Ellen Gleditch, 08.11.1914, National Library of Norway, letter collection.

[428] Letter from Birkeland to Ellen Gleditch, 12.11.1914, National Library of Norway, letter collection.

> diameter, like a Saturn ring. Can you guess what the dome will cover? When I am boasting I say to my friends, "Next to God I have the greatest vacuum-chamber in the world." I will make this idea a reality. I will make a vacuum chamber of 1000 cubic metres and, every Sunday, people will have the opportunity to see a "Saturn ring" of ten metres in diameter, "Sunspots" that no one can do better, and the "Zodiacal Light "just as evocative as the natural one, and finally Northern Lights around the polar regions of a terrella, which is 4 metre in diameter. The same sphere will serve as Saturn, the sun and the Earth, and will be driven around by an electric globe-mechanism, on the floor inside the vacuum chamber.[429]

In the dome he thinks there should be ten windows, which are of 50 millimetre thick glass so that people would be able to see through it, along a corridor surrounding the vacuum chamber, which should be made of thin bronze. The bolted granite dome will contain most of the air pressure. Evcrything is thought about in dctail. Finally hc says, "Onc could, havc used steel, which is as strong as the granite stone, but with so much iron or steel my phenomena would not look nearly as beautiful."

Egyptian girls standing to be photographed. Photo: Ole Andreas Krogness. Archive: National Library of Norway - Photo collection.

[429] Letter from Birkeland to Ellen Gleditch, 22.07.1916, National Library of Norway, letter collection.

He does not mention where in the world the museum will be established, or which mountain rock he has in his thoughts. Maybe it was a vision that included Haldde, in some way? The most beautiful heavenly phenomena people should have the pleasure of seeing in a museum. Birkeland's dream was not realized in his lifetime.

It is unknown when the Haldde residents received the news about Birkeland's death, or what kind of reaction they had. They probably wondered what had happened, but they make no comments. All that can be found in the letters about this is written by Ole to Dagny. He has been travelling for a while and among other things tells her that he has been in Kristiania in connection with, "the funeral ceremony for Birkeland". Ole wrote, "As you have probably seen in the newspapers I lay down a wreath from Haldde and Tromsø with some well-chosen words. You can now tell Devik that his telegram arrived and was referred to after Sæland's lecture in the evening."[430]

Constructing the market place. Photo: Ole Andreas Krogness. Archive: National Library of Norway – photo collection.

A short comment about this can also be found in Olaf Devik's book from 1971.

[430] Letter from Ole to Dagny, 24.09.1919.

> For those of us who knew Birkeland, the telegram that came from Tokyo about his death, was a message about a great personal loss. Everyone felt that now he has gone, it was one of the greatest losses that Norwegian Science has experienced in our time.[431]

In the newspaper, "*Aftenposten*" 16th June 1917, the Haldde researchers could read a long article that began:

> Professor Birkeland died Friday 15th June 1917 before he had reached his 50th year. It is an irreplaceable loss for Norwegian science. He was of a creative and constructive disposition. He worked with the most complicated problems, on the border between physics and mathematics. He was interested in the problems and secrets of the universe, and quite surprisingly for all of us, he found answers that scientists had been working with for centuries.

As Birkeland's closest working partners, Devik and Krogness, together with Birkeland's cousin attended the inheritance settlement, and were asked to examine all his books, manuscripts, letters and work-related documents. They were to be certain that no controversial documents, letters or instruments were lost, during the division of the property.[432]

It was never established what caused Birkeland's death. Neither do we know what the network of researchers, and his family, thought about his destiny. In the diary notes written by a Japanese researcher, T. Terada, who had the most contact with him while he was in Tokyo, suicide is mentioned as a possibility.[433] Not even in letters from Haldde or in Devik's book, or in other associated material that was collected, can be found any speculation about what happened to Birkeland in Egypt or Japan. A strange silence hangs over these texts. Egeland mentions in his biography about Birkeland that his death was overshadowed by suspicions of suicide. This silenced the praise he deserved from Norway, as a nation.[434]

Later it was known that Birkeland, according to Professor Alv Egeland, was known as one of the most original and independent scientifically talented people that Norway had ever had, but unfortunately during the last years of his life he had experienced big problems.[435] In the booklet, "*Kristian Birkeland, "Mennesket og forsker*", Egeland talks about

---

[431] Devik 1971:74.
[432] Egeland & Burke 2005:163.
[433] Egeland & Burke 2005:203.
[434] Egeland & Burke 2005:154.
[435] Egeland 1994.

Birkeland as a researcher and a person, "with a warm heart, and a loveable character, impulsive and easily enthusiastic. He was full of energy when ideas came rushing in over him or if he got new, exciting results". He could also be sensitive and somewhat bitter and easily irritated. "It was not only a few, who were envious of the fruits of Birkeland's work, and this envy brought forward quite a bit of opposition", Egeland said. He tells about Birkeland being found dead in his hotel room with a revolver and a bag of Veronal on the bedside table.[436] Veronal is a sleeping medicine, which Birkeland used to help his long-term insomnia.

Ole and Olaf talking seriously. Archive: Private, Mette Bunting

[436] Egeland 1994:50.

While Egeland, tells us that Birkeland felt he was being pursued and was paranoid about this, Jago, in her book hints that perhaps he was really being followed, and that he had grounds to feel his life was being threatened by enemies.[437] As earlier mentioned, it is clearly documented that Birkeland had some opposition within the scientific milieu, and that he was involved in industrial production with large capital investments.

The 4th April 1917, the day after he had sent the telegram about his arrival in Tromsø, he wrote a long letter to his near friend, Kaja Geelmuyden (1865–1918), who he had known since childhood. They were of the same age and had been together on social occasions and on travels.[438] In his letter he tells about the special experience he had in Egypt, and that he was really looking forward to seeing his friends in Kristiania.

> I long so extremely for Kristiania and my old friends. I came to Egypt in good faith, which is quite normal for me, with everything and everybody. For the first time in my life I have thoroughly learnt that one must not have blind faith in untried friends because they can, if luck will have it, deceive one more than the most dangerous enemies.

He does not disclose what the problem is, but stresses that he longs for home:

> [...] first of all to continue my work, of course, and next to write my own Odyssey, and place it in a secret archive. I have namely experienced things in Egypt this time that cannot be told to all and sundry. But I can also relate orally about my experiences to chosen faithful friends of the old school of whom I do not have many left. Together we can try to really figure out what has actually happened to me; first and foremost out of my own interest. I believe that Egypt is occupied by some of the world's most malignant-natured people. They always lie and deny facts point-blank even of things right in front of their noses. Now during the war they are worse than ever, because they use every means to lead the nervous and mistrustful Englishmen up the garden path, and they are easily led astray, and end up in real trouble.
>
> But I have the feeling that I have endured all the suffering for the sake of France, the country we both have loved since our childhood.[439]

---

437 Jago 2001.

438 Egeland & Burke 2005:161.

439 Letter from Birkeland to Kaja Geelmuyden, 04.04.1917, Norwegian version from Alv Egeland's archive 2014.

A mystical telegram from Birkeland to his lawyer, Johan Bredal, contributed to confusion surrounding his death. The telegram was sent from Tokyo a short while before he died, and said, "Please remember Mrs. Wriedt's committee". Bredal did not understand what Birkeland meant. It was also strange that Birkeland had written it in English, because Bredal knew very little English, and all the other telegrams Birkeland had sent him from Tokyo were written in Norwegian.

The reference to Mrs. Wriedt must be placed in an earlier encounter between her and Birkeland. Around the 20th century there had been a growing interest for spiritualism. While most people were sceptical to the possibility of communicating with dead spirits, some well-known scientists in England and America became members of spiritualist associations where mediums appeared. Mrs. Wriedt was a well-known medium. Through a hollow pipe, which gave out sounds and smells, she made contact with the dead. She came from America and had in August 1912 a séance for three evenings in St Olaf's Hotel in Kristiania. Birkeland was asked to lead a scientific committee to give an objective evaluation of the woman's spiritual ability. The committee was present for the three evenings, and concluded that it was just cheating and swindle. Birkeland unveiled that the pipe that hopped and jumped when the spirit spoke was covered in a special stuff. This resulted in small explosions and it was no more than a clever magician could do. Kaja Geemuyden was also there on that occasion. Both she and Birkeland reported the event in the newspaper, "*Tidens Tegn*" 18th August 1912. While she described what the voices from the pipe said, and how the participants recognised their own dead relatives and other acquaintances, Birkeland gave an account about how he had revealed that Mrs. Wriedt was a fraud by investigating the contents of the pipe. His comment in the newspaper about this, was:

> It is my complete conviction that Mrs. Wriedt is an incredible cunning humbug swindler, who has run riot for about 20 years. If we allowed ourselves to be blinded by such, we could easily return to the middle ages again. […] I am certainly principally against all burning of witches, but, even so, a little bit of fire in honour of Mrs. Wriedt would not have been misplaced.[440]

[440] About this occasion see "*Tidens Tegn*", Sunday18th August 1912, page 6. "The spiritual humbug. The control committee disclosure." Professor Birkeland's declaration. "An amazing humbug maker." A joint statement by the committee. "Any magician could do the same." The women's declaration. An accurate report of the first séance. Kaja Geelmuyden wrote this last report.

The newspaper press in Kristiania reported a statement from the audience in connection with the séance:

> Birkeland and the others can talk about humbug as much as they like. It is not something physical that is the question. If there had been a slight amount of explosives in the pipe, so what? What the voices tell us nobody can explain away. Today we shall make a chain of love. It is love that does it. We are crazy with happiness.[441]

It seems most unlikely that Birkeland should have changed his view about this and almost encouraged his lawyer that after his death he should make contact with him through a spiritualist. But why was the telegram sent and was it sent by him? He had informed the Haldde researchers that he would be coming in connection with the planning of a new institute in Tromsø, and he had told his close friend Kaja how much he yearns to be home. He wanted the two of them to try and find out what had happened in Egypt.

The telegram led to Sir Oliver Lodge being contacted, He was a British physicist with an interest for paranormal phenomenon. He was asked to find a medium, who could try and communicate with the Professor. They managed this, but their attempt did not work. Birkeland's spirit remained silent.[442]

The staff from the Norwegian Embassy in Tokyo packed Birkeland's scientific documents to send them with the Swedish steamer, "Peking" to Norway. At the end of August 1917 the ship sailed from Japan. A wireless communication from the ship was received 2nd September 1917, as it got close to Korea. Then it disappeared, and in October was listed by Lloyds as missing. This is how most of Birkeland's belongings and documents disappeared, and perhaps they may have been able to shed some light on what he had experienced of difficulties while living in Helouan. Not even this is mentioned in the letters from Haldde or in Devik's book. The fate of the researcher, who founded the Northern Lights' Observatory on the summit of a mountain in Finnmark, remains a mystery.

A speech to celebrate Birkeland on his 50th birthday 13th December 1917 had been planned. Parallel with the preparation of this, six well-known people within research and industry had begun to work on an

---

[441] Hølaas 1945, from year 1912.

[442] Jago 2001:222. We have not found any documentation that there was a séance. It is however documented that Olaf Devik received a letter dated 9th November 1967 from a legal assistant to Bredal, named Henning Bødtker, with some information about the telegram. (Egeland & Burke 2005:156).

application for a Nobel Prize in Physics for Birkeland.[443] They had made contact with the Nobel foundation to get scientific support and advice. Then the telegram arrived 17th June 1917 from Japan about Birkeland's death, "and the plans for a Nobel Prize went up in smoke", as is written in a letter dated at the end of June.[444]

Birkeland had been nominated for a Nobel Prize for chemistry four times earlier in 1907,1909 and 1913, together with Sam Eyde, and in 1912 on his own. The nomination in 1912 angered Eyde, who meant that it was with his money that Birkeland had been able to invent the furnace, which could produce artificial fertilizer. This incident increased the conflict level between the two earlier cooperative partners. It is not known whether Birkeland knew about the nomination that was being planned in 1917.[445]

It was first after the war that Birkeland's ashes were carried home and buried, as we found out through Ole's letter to Dagny. "He was a researcher, who was buried with the Grace of God," according to the commemorative speech. On his gravestone is written, "He bound nitrogen from the air with an electromagnetic arc. He studied the Northern Lights, solar radiation and the earth's magnetic field."[446]

In October 1917 a fond was established, "Professor Birkeland's Fond for Geophysical Research". Many were responsible for supporting this, among them; the Prime Minister and the President of Parliament, most of the professors at the Scientific faculties at Norway's two Universities, President of the Norwegian Academy of Science and Letters, a number of industrial leaders, well-known authors and legal practitioners. The rent from the collected fund was to be used to stimulate geophysical research, especially in Northern Norway.[447]

---

443 Egeland 1994:51.
444 Egeland 1994:51.
445 Egeland & Burke 2005:139.
446 Devik 1976:76.
447 Egeland & Burke2005: 175.

# CHAPTER TWENTY ONE

# FACING DEATH

There was not very much that could be used as help against the Spanish flu. For Dagny brandy was the best medicine from the chemist, "*Nordstjernen*", shown here in Tromsø. Archive: Tromsø Museum - University Museum

*Yes it is a terrible illness; here they are dying around us.*[448]

The Haldde family's arrival in Tromsø was similar to their arrival in Kåfjord, full of drama. It was 1918 and the Spanish flu had reached Tromsø. This was the second and most serious wave of influenza, which had hit the country from north to south. One believed it was probably spread through transport, such as boats, train and car. Dagny writes in her memoirs, "So the Spanish flu arrived and it is terribly devastating. A great many young people have died. We have a prohibition in the country so

[448] Letter from Dagny to her mother-in-law. 21.09.1918.

that no one can get brandy, which we regard as medicine. We have to lure it from the chemists.[449]

It is not the lack of brandy that is the most precarious in this situation. In Tromsø like other places in the north, it is the shortage of doctors.

The doctor often had to travel a long way to visit people who were ill in small places outside town. Photo: Thoralf Øien. Archive: Perspective Museum, Tromsø.

World War 1 is still raging, so it is difficult to get access to doctors, and others who can help sick civilians. The doctors have to take care of the soldiers, also those that are not active in the war, but who are collected together here in this region. Dagny tells her mother-in-law that; "[…] Doctor Holmboe who is 75 years old is out every night in a motor boat, on long sea journeys. The military have called out the hospital doctors, so that the ones that are left are overworked".[450]

Dagny's concern for Doctor Holmboe proves to be justified. We can read in the history of Tromsø that the ageing county-doctor, Johannes Holmboe, worked 14 to 15 hours every day, including Sundays. He managed with just a few hours of sleep, right up until he died unexpectedly while

[449] Memoirs for her grandchildren written by Dagny, 1976.
[450] Letter from Dagny to her mother-in-law, 21.09.1918.

working. Together with other doctors he had fought a terrible fight against illness, in the summer of 1918:

> One of the last days before he died he had been out in the district during full winter weather, and visited thirteen farms with forty people who were ill, and got back to town at 3 a.m. Early next morning he sat in his office, with a waiting room crammed full of sick patients.[451]

On his way out on another visit, the boat he was on, tragically capsized and the exhausted doctor Holmboe fell into the water together with four other men. While people on a local boat saved the others, the doctor sank and was never found.[452]

About half of Norway's population suffered from Spanish flu, while it raged across the country, and between 13 000 and 15 000 people died in one year. The majority of these had complications from pneumonia that developed after having Spanish flu. The influenza came from Spain and spread along the military front and into the trenches in the war ravaged Europe, and travelled further north with the coastal boats. Around 25th July the disease spread, "like wild fire", over the town until it reached its culmination in September and October, and then it was over.[453] During that period the life expectancy for both sexes sank, with seven years. For men it became 48.5 years of age, and 52 for women. Researchers in later years presume it was a form of bird flu virus that spread so quickly over the whole world.[454] Dagny is witness to what it was like to catch it:

> There are many sad instances; mothers with 6 to 8 children died of the Spanish. Just in one house a wife and two daughters died and in another place all three of the parents' children died. Yes we can be glad to have been spared. I do not know if cleanliness has anything to do with it. There are four people with it living wall-to-wall to us in the other courtyard, but no deaths, the one young girl had pneumonia, and the doctor gave her camphor and brandy.[455]

---

451 Ytreberg 1971:41.
452 Ytreberg 1971.
453 Ytreberg 1971:47.
454 Borza 2001.
455 Letter from Dagny to her mother-in-law, 27.10.1918.

It was new for Dagny that she could walk in Tromsø's streets and not least of all to be able to buy goods directly from shops. Perspective Museum, Tromsø.

Those who were infected often took the infection to those nearest to them, and visited such places where many people were collected, for example in school, at the market, in church and not least at many funerals. It was from droplets that one was infected, and it took two days before the first symptoms came. These were similar to how we experience influenza today, just immensely worse. There was a sudden high temperature, pain in one's eyes, ears and back, tenderness in one's head and pharynx, bleeding from the nose and dizziness. Those that were infected were ill from a few days to a week, but could experience other complications afterwards, for example pneumonia, which many died of.[456]

[456] Borza 2001.

The view from the top of the Island of Troms, "*Tromsøya*". The brick building on the left is Kongsbakken, Tromsø's State School. It is on the way up to the Geophysical Institute. Photo: Thoralf Øien. Archive: Perspective Museum.

There was little sickness on Haldde Mountain, but after they had come down to Tromsø, they catch colds more than ever. Then their own household is hit by the Spanish flu, as Olga, their maid, becomes ill. Suddenly the roles in the house are turned upside down. From having a girl to help her in the almost finished flat in the Geophysical Institute, Dagny now has a young patient who needs to be looked after, and given every attention. Dagny is pregnant with her third child, and she is once again beginning to feel that she does not have much strength left.

Dagny helps the young girl and looks after her as well as she can. Although Olga becomes even more ill, and it becomes even more difficult to take care of her properly. "She lay in bed for 12 days, but when the night vigil began I felt that I just could not cope, and there was no one there who could help me."[457] Dagny manages to get some help from Olaf Devik's maidservant, who dares to watch over Olga for one night, but she has to get back to her own duties. A solution is to leave the door open into the family's bedroom. Dagny cannot cope with being continually woken

[457] Letter from Dagny to her mother-in-law, 21.09.1918.

through the night, while at the same time looking after her own children and house in the daytime. In the end she gets the sick girl to the hospital. "I got her to the hospital on Monday, as she had to have injections, but she died in the night at 2 a.m. of heart failure. She had breathing problems the last two days otherwise no energy and was very weak."[458]

Dagny also finds out that it was not directly Spanish flu that killed the girl, but complications afterwards, just like so many others. "She got pneumonia and had heart failure, apparently she had pneumonia the last two days, it is such a deceitful illness. She was not particularly strong and did not have much resistance to illness."[459] Ole was away during the time the girl was ill, and Dagny tells her mother-in-law that Ole met the girl's coffin on the boat when he was on his way home.

Dagny admits in her letter that she is nervous about being infected, something she was not alone in thinking at that time. When Synnøve also becomes ill, she quite naturally is afraid that it is Spanish flu. Luckily it turns out to be bronchitis. Most of her letters from this period to her mother-in-law begin with an assurance that they are not infected.

Throughout the whole of Olga's illness she has been extremely careful with cleanliness.

> I have been so careful about washing my fingers every time, after I touched her, and I boiled all the cutlery and dishes so I think the infection is over, but it chooses one here and one there. It has made such an impression on me, and Ole thinks it was sad.[460]

After the girl's death Dagny writes to her mother-in-law. "We were over yesterday to get some bedclothes for the new girl, as Olga's must be disinfected. […] I hope sincerely that we do not get it."[461]

As the quotation above shows, Dagny soon employed a new girl, in fact the day after Olga died. Dagny had asked Mrs Devik to advertise in the newspaper, "after temporary help, as I had to have help, because Olga would not be able to work for a while. Then a young girl turned up, who looked pleasant, and I made an immediate decision."[462] As things turned out as they did, with Olga, Dagny was satisfied that she had made the difficult decision of finding another girl to replace her, even though Olga had still been alive at that time.

[458] Letter from Dagny to her mother-in-law, 21.09.1918.
[459] Letter from Dagny to her mother-in-law, 21.09.1918.
[460] Letter from Dagny to her mother-in-law, 21.09.1918
[461] Letter from Dagny to her mother-in-law, 21.09.1918.
[462] Letter from Dagny to her mother-in-law, 21.09.1918.

Despite Dagny's focus on cleanliness she was eventually infected. The letters we have access to do not completely tell us when it was, but she mentions it in her memoirs. "I got the Spanish flu, but managed to get over it. Your grandfather sat on the edge of the bed and gave me warm toddies (the chemist's brandy) and I also had medicine, and Dagny Devik made me oat-meal soup, - so I was meant to live."[463] Luckily all went well with Dagny, and she gave birth to a daughter in April 1919, Inger Augusta, her third child.

[463] Memoirs for her grandchildren written by Dagny, 1976.

# CHAPTER TWENTY TWO

# FROM MOUNTAIN TOP TO CITY LIFE, AND A POSSIBLE UNIVERSITY?

Researchers and wives gathered together on the Geophysical Institute roof terrace. Archive: The World Heritage Rock -Art Centre, Alta Museum.

*Here there are just lamps and sadness in the evening. I just hope Ole does not travel away from me again.*[464]

Despite everything, there is quite a lot that indicates that life on the top of a mountain was easier than it was in Tromsø. One concern was the tragic circumstances surrounding Olga, their maid, who had moved with them, became ill and died, and that the Spanish flu caused havoc everywhere and took many lives. Another concern was the war and

[464] Letter from Dagny to her mother-in-law, 27.09.1918.

rationing of food. Otherwise it is clear that the earlier organization of scientific work and daily life had formed a perfect union in the collective research colony. This had, in many ways, been better for the women than what they came to in Tromsø, at least in the beginning,

While the house for the Geophysical Institute and the families' flats were being built, the families from Haldde were lodged in temporary housing in Tromsdalen, with Tromsø Sound between them and the city centre on the island. To get there they had to take a ferry. Ole had an office on the island in Tromsø Museum, and left early in the morning on the ferry, which meant that Dagny had no other choice than to spend her day on the mainland. She felt there were many drawbacks with this situation. The worst was that it was almost impossible to buy any food on that side of the sound where they lived. Her letters to her mother-in-law, in autumn 1918, contain many complaints. "It is somewhat difficult to live here because I cannot store any food. I am not used to buying food for each day and think it is an unnecessary waste of time to have to travel every other day to get food."[465]

Dagny was used to having the food frozen in a snowdrift outside the house. This was impossible in a town, as the snow was not clean enough. She had to get someone she knew to do the shopping for her when she was unable to do it. Food was not delivered to her as it had been on Haldde. Also, it was not always possible to get hold of the food she wanted, "Yes, let's just hope that rationing will soon stop and then we can get cheese and eggs and all the other things again. At the same time it is bad of me to complain when things are so terrible for people in Germany."[466]

That Mrs Devik was no longer in the same house, and could help Dagny to look after the children, also made things more difficult. Not even Mrs Devik was content with her existence according to Dagny:

> The Deviks' had to move to Dybwad Holmboe's summerhouse yesterday. Tulla has got an ear infection and they cannot take her over the sound again. They had the child in the kitchen in a playpen, which stood between two doors, so she must have been in a draught. Mrs Devik has been going to town quite often because she does not like being at home on her own without her husband. Unfortunately she has a maid who runs away from the child, she has often been here and left the child alone, so I do not think it is strange that things ended up as they did, but the flat will get the blame. I will have to tell Mrs Devik, because it is so dreadful.

[465] Letter from Dagny to her mother-in-law, 17.10.1918.
[466] Letter from Dagny to her mother-in-law, 17.11.1918.

Once again Dagny has trouble with her maidservant:

> The new girl I have is as slow as a louse on a stick covered in tar, but she has never been a maidservant in a house before. She has both county school education and domestic science courses, but the worst is that she does not have a sense of order. I have thought of trying to have her with us to the Geophysical Institute, because in the flat there will be more room for things than here. Maybe it is good for me, as I have always had girls who have been responsible for tidying up themselves[467]

It was also a readjustment for Dagny to get to grips with the busy town life. That everything in the town happened at a different tempo than on the mountain was shown in different ways. For example, the post-dispatch no longer came with the reindeer and sledge up the snow-clad mountainside. Parcel post from her mother-in-law was handled completely differently and more brutally, "Do you know what, the salt bowl was in a hundred pieces! They throw the post so recklessly here, it is so sad but there is nothing we can do about it."[468]

Dagny had her birthday in the autumn. At the research colony on Haldde, Dagny was used to having a small birthday party, but things did not go quite as she planned in Tromsdalen:

> For my birthday I only received a letter from you and no parcel, it was namely an unlucky day. First of all none of those I had invited could come, all of them were busy. We actually have a sewing group, but I decided I could not invite them too early, because I was unsure if I could get anything to offer them, as we have very little here at the moment.[469]

It also appears to be a problem for those who were not busy; they preferred to come in the afternoon. Dagny had thought of having the party after the children had been put to bed, but not even that was possible:

> We would have to hire a boatman after 9 p.m. because the ferry does not go after. 8.30 p.m. Therefore it is of no use having anyone over here in the evening. Just before supper Ole had to go to town for Devik to fetch a doctor, because Tulla had a temperature of nearly 41°C. It is so difficult to get a doctor here, because of that disgusting ferry.

---

[467] Letter from Dagny to her mother-in-law, 17.11.1918.
[468] Letter from Dagny to her mother-in-law, 17.11.1918.
[469] Letter from Dagny to her mother-in-law, 17.11.1918.

After 9 p.m. an extra ferry over Tromsø Sound had to be ordered to row between the town and the mainland. Here is the boatman, Kiil, rowing. Photo: Boye Strøm. Archive: Sparebanken in Northern Norway, Tromsø.

It was not surprising that Dagny disliked being dependent on a ferry. The ferry connection over the Sound started in 1914, first of all with a small ferry called "*Lilleferj*a", It was painted black and people in the town soon re-christened it to "*Likkista*", (coffin), mainly because they experienced, "life-threatening confinement", cooped up under the deck with no emergency exit if anything happened. The same year a new bigger ferry, "*Storferja*", was put into service, which had some better emergency regulations on board.[470] Even so transport was difficult from Tromsdalen to the centre of town most of the day, when the ferry did not go, and the boatman had to be contacted.

[470] Ytreberg 1962:83.

The "disgusting ferry" which was colloquially christened "*Likkista*", according to the owner of the photo. Here is the boat in a reconstructed version from 1930. Archive: Private, Reidun Tøllefsen.

Dagny writes that she worries about Spanish flu. It takes a lot out of you having so much sickness all around, and they live wall-to-wall with a family where most of them have been ill. Dagny is afraid of being infected. She is also irritated with her maid. That she herself is pregnant again does not make things better. The third child is on its way. Dagny feels listless, unable to do any work and incapacitated. She regards herself as more irritable than earlier. She is unable to tolerate standing for very long, and thinks she has so many disadvantages. Both her physical condition and her own sense of well-being, bother her. Her appetite is also bad. "There is no food to get here that helps, so I do not have a large appetite and one does not grow just on whale fat."[471]

The person, who is enjoying their new life situation, is Ole. He has quickly joined the town's music milieu. Dagny does not participate, she tells her mother-in-law:

> This morning there is a concert in town. Ole is playing in it. I have got tickets three times, but have been unable to go because of Synnøve and the ferry. I have nothing against this as long as she gets better, but it is rather monotonous to always be sitting beside a sickbed.[472]

The music life in town gives Ole the opportunity to really cultivate his love of music. Also in connection with his work situation he is often invited to various arrangements in town. To begin with Dagny had a

---

[471] Letter from Dagny to her mother-in-law, 17.11.1918.
[472] Letter from Dagny to her mother-in-law, 17.11.1918.

position on the sidelines, something quite different from her life on Haldde, where there were not such big parties, other than those they initiated themselves. As far as her mother-in-law was concerned all was well. "Ole was at the Mayor's dinner party at, Magistrate Meinich's, who appears to be held in high esteem here in town, and there were many amusing songs written about him. I thought it was fun to smarten Ole up with a bow tie."[473]

Not completely pleased? Photo: Ole Andreas Krogness. Archive: National Library of Norway, photo collection.

We do not know whether things were as fine as the letters imply. Ole uses the social situation to help him make useful contacts. He often aroused interest for his bold actions when conversing at social occasions. A comment from Dagny was, "Ole often works hard at social

[473] Letter from Dagny to her mother-in-law, 11.01.1919.

occasions".[474] But there is a photo of a furious Dagny taken in Tromsø. Maybe it can tell us about controversies which mother-in-law has not been told about?

The climate is also different than on Haldde. It is the middle of November and still no snow. The weather alternates between rain and snow. It is icy and slippery. Dagny has obtained a walking stick, which she and Ole share. Now and again she goes for a walk with a friend, who is also pregnant, "[…] she is expecting too, so it is an imposing sight to see us together, her delivery date is a fortnight after mine, […] but she thinks she is so much slimmer than me."

After they moved into their new home, in the Geophysical Institute, right on the top of Troms Island, they arranged a musical evening in their own flat. Dagny's mood improved considerably. Now Ole was closer every day, and their social life improved even though they were in the middle of an expensive time after the war. It was almost impossible to get food and drink for such social gatherings. Others were better off than research families.

> I do not know how we are going to manage a large party. We have been to three big social occasions, but we cannot possibly have a return party, as we need both food and much more. There is no wine to be found, but they use fruit wine. We were at the Chemist Dahle and there we got some white wine, red wine and dessert wine plus beer. So chemists live a life of happy days.[475]

In the middle of this somewhat dejected contemplation about everyday trivialities, the idea of a university is further developed. Only one critical voice had been raised. That was the North Norwegian fisheries inspector, Oscar Sund, who said he thought it was rather a wild idea to talk about a university for the region, when the real cultural need was for all kinds of schools.[476]

The Geophysical Institute was now a reality, and this laid the foundation for continuation. During the winter of 1918–1919 the case for establishing a university was discussed by scientists both in the institute and in the Museum, which until the institute had been established was the only scientific institution in the northern region.[477] Now the idea was to unite the two institutions.

---

[474] Devik 1976:115.

[475] Letter from Dagny to her mother-in-law, 11.01.1919.

[476] Devik 1971:97.

[477] The National Archives, Private archive 686, Olaf Devik, number 11.1.

Conservator, Carl Dons at Tromsø Museum, had advocated the constitution of a, "Scientific Company for Northern Norway", or an "Arctic Faculty". Consul Conrad Holmboe gave 100.000 kroner to Tromsø Municipality, to further the planning of a university, and he requested that plans for a realistic goal should be made first of all. A work programme was set up for long-term planning The same autumn that the Haldde researchers came to Tromsø, Krogness, Dons and Devik wrote a report with the title, "The first steps towards a North Norwegian University: The establishment of a Grant Fund for investigation and scholarship, and a Scientific Academy in Tromsø."[478]

[478] See Devik 1976: 96-103.

The music life in Tromsø was an Eldorado for Ole. A lot took place in Alfheim, the city's popular culture centre with an open-air stage, but also in other places music was played to the great delight of the public. The music pavilion in the middle of the city centre was a meeting-place. Photo: Thoralf Øien. Archive: Perspective Museum, Tromsø

Tromsø Museum, where Ole had his office while the "*Geofysen*" – was being built. Archive: Tromsø Museum – University museum.

They pointed out that the region had both developed and progressed greatly, and now they wanted to create, "a cultural and scientific centre" in the north. They believed that a scientific college would help improve the lack of government officials in the northern region, if the youth out in the districts areas could more easily obtain a higher education. They also felt that at long last one was now aware that in Northern Norway there was, "a productive and a unique area for scientific research".[479]

The group had drawn up a rough estimate of the expenses that they would need for job positions, wages, a new building and running costs. To decide how the funds should be used, they thought that a committee should be established with representatives from Northern Norway's different regions.

Provisionally they aimed at establishing a Tromsø Academy.[480] Interest in marine research and the practical interests in connection with the fisheries made it obvious that the work should begin as soon as possible, Devik writes in his book. A main question was:

[479] Fulsås 1993:113.
[480] Devik1971: 97.

> Is it possible in the near future to create an institution, that both will have scientific significance and moreover realize, at any rate, part of the practical purpose from which the idea of a university has sprung?[481]

What they wanted was to make Tromsø a centre of scientific development for Northern Norway.

In May 1919, Parliament passed a resolution to given an additional grant to the Geophysical Institute of 60.000 kroner. This extra encouragement showed that there was support for local enterprises. The Cabinet Minister Løvland encouraged "rich men" in Tromsø to give money to a university fund, even though it could take a long time before the university could be established.[482]

The report that Ole and Olaf had written together with Dons was sent with an invitation to specifically chosen town citizens. The invitation sounded like this:

> With reference to the enclosed circular produced by leaders from the Geophysical Institute and the Museum 's assistant curator, we permit ourselves to invite you to a meeting in, "*Sparebankens*" Assembly Hall, Saturday 5$^{th}$ July at 9 p.m. to discuss the question of the establishment of a Scientific Academy for Northern Norway, and what this entails. We allow ourselves to request that you make this meeting known within your circle of acquaintances, as everyone is welcome to participate.[483]

The meeting was held on a fine summer's evening with sunshine and warmth. Quite a number of people met in the bank's Assembly Hall. Speeches were held by, Rector Quigstad, Hans A. Meyer, Headmaster Grønlie and Ole Krogness. There was a general agreement to the plan and a resolution was passed:

> The participants at this meeting give thanks for the idea of establishing a study fund and a Scientific Academy for Northern Norway, and give this their warmest support. It must be regarded as a fully legitimate demand that this large region of the country should also have its own higher cultural institution, in the same way as Østland, Vestland and Trøndelag.[484]

A working committee was voted in, with representatives from all three northern counties. The perserverent Meyer from Mo i Rana, was completely

---

481 Devik 1971:97.
482 Fulsås 1993:13.
483 Devik 1971: 99.
484 Devik 1971:100.

clear that the aim should be a university. Even if the first plan was to be modest, it was, as far as he was concerned, a case of, "being alert, keeping options open and being prepared for direct action, when the time comes."[485]

Work with the plans for the university continued, parallel with the tasks at the new institute, into which they had moved for Christmas. Once again the scientific and daily work was amassed under one roof. Dagny is pleased with this situation. She tells her mother-in-law that they will be appointing a new employee, who will both help in the house and with the children. Miss Tveten is to be paid by the institute for office work, and the family is to supply her meals. They cannot afford to pay for private help. Even so, a number of everyday troubles are not over yet. Dagny takes things with good humour:

> We will probably get," *flektyfus*" soon, as here in the north louse swarm in every house. We have now got rid of them, my previous maid had a head full of them, I do not know if I wrote about the louse problem over Christmas. I almost had to cut off all of Oa's hair, but he did not catch a cold.[486]

During Ole's visit to Kristiania in connection with the commemorations for Professor Birkeland, it was clear that he is unable to go straight back home again, as he really wanted to, and assist Dagny with her hard endeavours. After a number of different geophysical meetings he has to go to a conference about Polar investigations in Paris, as a representative of Norway.

---

[485] Fulsås 1993:15.

[486] Letter from Dagny to her mother-in-law 11.01.1919

"*Geofysen*" – the Geophysical Institute was ready from November 1918. Photo: Thoralf Øien. Archive Perspective Museum

While Ole is in Bergen waiting for a boat connection over the North Sea, he makes use of the time to enjoy himself while at the same time working hard. To his, "Dear little wife" he writes, "So yesterday I was out equipping myself. […] I got 50% price reduction to Newcastle and it was disgraceful to appear in my old suit among the distinguished people, who we will be meeting in Paris."[487]

In the same letter he is concerned about helping Dagny find a good maidservant. He has been in contact with a clergyman's daughter from Sogn, whom he has written a letter to, and asked her to telephone or preferably to travel to Bergen herself to come to an agreement about the terms for the domestic position. "It is probably best that I bring her with me when I return. She is a pleasant and modest kind of person, and everything will be fine, just wait and see."[488]

He has also been to some friends and experienced that they also have problems in getting a maid. They have been without help for half a year, and Ole wonders if Dagny can help them:

[487] Letter from Ole to Dagny, 16.09.1919.
[488] Letter from Ole to Dagny 16.09.1919

> Don't you have a reliable girl up your sleeve, one you can recommend, and one that can make ordinary food and scrub the floor? If so you must help her. Perhaps you could write to her, (Mrs Helland-Hansen) address, Hop pr. Bergen, if you can find someone for her. It is quite normal for us in Tromsø to get people here from Bergen, and that people in Bergen get someone from Tromsø(!) There are 5 exceptionally sweet children, and the couple are also very pleasant. They have just appointed a woman as housekeeper.

Ole is mostly concerned with the journey he is about to take. Before Bergen he was in Kristiania and had some profitable conversations with the governmental department that worked with their case:

> Løvland was a bit worried about sending all three men as everything is so expensive, but I managed to talk to him 23rd and persuaded him to change his mind. When he heard that the Birkeland Fond would also give some support, then the Government could manage the rest. As far as I am concerned I would have preferred to return home, but there were so many practical difficulties in connection with the Polar investigations, that it was difficult to get out of. Hope that all is well with you up there, and that you have got some help. My address in Paris is most likely the Hotel Sylvia Boulevard St Germain 51, but as the post takes up to a fortnight to slip through the censor, you can send me a telegram about how things are.

While he was in Kristiania, he also visited his mother, who in her letter to her daughter-in-law, mentions the visit

> I heard that he was going to Birkeland's funeral that day, to put a wreath down, and then a meeting with the Government and much more, and was very busy, as he was planning to travel to Bergen again that evening. He borrowed Jens' top hat, overcoat as well as his gloves, and as you can imagine everything went as quickly as possible.

Ole has apparently not been satisfied with travelling in the borrowed clothes that his fussy mother had palmed off on him. Perhaps the travelling companions, his mother tells her daughter-in-law about, inspired him:

> He rang the next day from Hesselberg's, that he was not going to leave until the following day, and so he was at your mother's that night, and that was good, because from what I hear there is nowhere to stay here. So he left the next evening with Hesselberg, Helland Hansen and Professor Schettelig for Bergen; so he had good company.

As usual Ole's mother is pretty well kept up to date about her son's activities in connection with his work, and the research milieu he is a part of. However, she is also concerned that he has a wife and children to look after, which he must not forget.

> He would rather have travelled home if he had been able to follow his own wishes, and I am witness to that, but he feels that it was a duty that was placed on his shoulders. [...] I really hope that you get a clever and decent girl and that Miss Østby can manage to do this and that, and understands that she needs to get on well with the children.[489]

Dagny had caring people in her near family. Her everyday was hard, but was full of resources and fellowship around trivialities. In the midst of this network of connections between people, who stood more or less close to each other, larger plans developed. Research had born fruit through determined work. Ideas were realized and new seeds were sown. A university had started to germinate.

[489] Letter from mother-in-law to Dagny, 27.09.1919.

# CHAPTER TWENTY THREE

# GENDER, CULTURE, CLASS

A regular supply of reasonably current newspapers on Haldde made it possible to be kept up-to-date on news items. Archive: National Library of Norway - Photo collection.

*[...] There has been a storm now for eight days, but happily today the reindeer was able to come up. We bought it in Bossekop. It is expensive but very gentle, and looks handsome with its big horns.*[490]
*[...] have a Sami boy who is very kind and willing. He chops wood and he is going to bring the mail up to us during the winter.*[491]

The main theme of the Haldde research material is woven into a North Norwegian cultural background. It is a unique region that contains several

[490] Letter from Dagny to mother-in-law 29.11.1912.
[491] Letter from Dagny to mother-in-law 11.10.1912.

ethnic cultures, Sami, Kven and Norwegian in addition to several new immigrant cultures. The cultural concept contains thoughts, ideas and traditions, which arise out of values and practices based on a variety of different experiences and role-models, of those who live in different cultures.[492]

The cultural development between the ethnical groups has not been on equal terms. Originally the Sami indigenous culture constituted the majority in this region. The concepts, minority and majority cultures, are terms which denote the power aspects of cultural development in this area. The Norwegian majority society has, through later history, been a dominating factor in Northern Norway, but the social hierarchies exist both within and across these cultures.

The letters contain traces of cultural encounters between the south and the north of the country. In their contents we also find gender cultures, which grew out of the class system, and which were kept in place by socially structured regulations. Working people, and the middle class, each representing their social class with different basic values and cultural opinions. It functioned according to the working patterns, and technological development of that period of time. The work done by women and men often supplemented each other, but the balance of power was usually unevenly distributed, both between and within the gender cultures. It makes sense to use concepts like working-class and middle-class when we are preoccupied with everyday life and household work.

In addition to the hierarchical order within the household between housewives and maids, the actual difference in power was manifested through the differences between the status and social position of women and men. If we regard the activities on Haldde through a gender perspective, by referring to some of the contemporary central female characters, who were critical to the existing social conditions of that time, then interesting connections appear in our research material. We will take a closer look at some outstanding women pioneers and what they stood for. They were active during the same period as the Haldde material is taken from. Through these characters we will throw some light on the cultural contexts that gave different living conditions, from which everyday life's actions originate; actions which at different speeds lead to changes in major social structures.

It was a woman who first put Kåfjord on the map of international activity. The young Marit from Dullan found copper ore in the mountain above the village. When she handed over the result of her discovery to the

[492] Klausen 1992.

shopkeeper Klerck, she also lay down the basis for the founding and establishing of a large-scale mining enterprise, which attracted both foreign capital and scientific research expeditions interested in the Northern Lights.

Marit was of Kven ancestry. She herself came from a generation of small farmers, who had emigrated from Finland and settled in Kåfjord.[493] The mining industry opened up different opportunities for other businesses, than just fishing and farming, and it attracted workers, trades people and governmental officials from nearby areas. A class society grew as a result of the Copper Works. This development, which the mining industry brought with it, weakened the tradtional Sami community in the northern region.

A Sami pioneer woman reacted to what she experienced as a disruption of her own culture. Else Laula Renberg (1877–1931) established in 1904 "*Lapparnas Centralforbund*" which was the first historical Sami association.[494] She travelled round in the Sami areas of Norway from Trondheim in the south to Finnmark in the north and gave lectures about the Sami situation. She was concerned about basic foundation rights for the preservation of the Sami culture, their right to the land, water, and formal education. She especially encouraged Sami women to organize themselves and use their own influential power to remove social injustices.

In 1910 she initiated "*Brurskankens Samiske Kvindeforening*", an association for Sami women, and she was its chairperson.[495] In 1916 she applied and was given Governmental subsidy to organize a Nordic political Sami meeting in Trondhiem, which was the first of its kind. The organizing committee only consisted of women. They encouraged all women to participate in the meeting and work for the Sami people. The meeting was reported in the womens' magazine *Nylænde*, founded by Gina Krog, a feminist pioneer, in 1887. In it was written, "Emancipation thoughts in their highest form, social awareness, has reached our sisters in the mountains."[496] This happened at the time when Dagny and Ole lived on Haldde. We can find no connection to Sami culture in our material other than Dagny's visit to the Sami reindeer husbandry market in Bossekop and that the Haldde residents obtained a reindeer with a driver to help with any necessary transport.

---

493 Pedersen 2008:14-62.
494 Skard 2013:28.
495 Hætta 1992:58.
496 See M.Hj., in periodical Nylænde 1917:87.

Schoolboys on their way to their future. Archive: National Library of Norway – photo collection.

Ole's appointment as Director of the Research Observatory on Haldde, brought with it certain responsibilities and expectations with regard to Dagny's status as a housewife. That a special type of woman was needed to fill this role is witnessed in the letters Dagny writes to her mother-in-law. Many of the letters are mainly concerned with a housewife's work. There is nothing that shows that she questions what tasks she is expected to carry out as a housewife. She accepts that it is her responsibility to see that everything is as it should be so that her husband and his research assistant can do a good job. Were there, even so, any signs which pointed towards Dagny feeling that a woman's role was beginning to change?

> It is touching that you want to make such an effort and knit him some socks, it ought to be my duty now, but I take it that it is something you love doing. It is quite right that it will be very beneficial for Ole to have more socks, because he only has two pairs and has to change them once he has been outside.[497]

[497] Letter from Dagny to her mother-in-law 29.11.1912.

This is written in one of the first letters Dagny writes to her mother-in-law. She is grateful that the mother-in-law relieves her from some of her duties as a housewife. At the same time she is absolutely clear about what her mother-in-law expects of her son's wife, but smart enough to appeal to her mother-in-law's need to be able to continue her loving care of her son.

As the wife of a researcher, Dagny had some standards that she had to live up to. Devik mentions in his book how the young wife to Vilhelm Bjerknes, son of the mathematician Professor Carl Anton Bjerknes, followed her mother-in-law's traditions. She looked after the house and financial accounts, in the same way as Dagny did on the top of Haldde, as well as being hospitable towards the many students and co-workers that came to their home.[498] Devik met this mother-in-law, a lively, wise old lady, who he felt must have been, "a good and practical providence for her absent-minded and impractical husband."[499] Vilhelm Bjerknes also mentioned his mother, in the biography he wrote about his father, and Devik concludes that it is justifiable that the wives of researchers are at times given some honour for their contribution. He never mentions anything about the maidservants' contribution.

Dagny moves to Haldde with her husband during a time of upheaval in the history of Norway. A year after they arrive on the mountain, Dagny together with other women in the country, was given the right to vote for the National Assembly, the country's supreme democratic organ. Dagny does not mention this in her letters. When we, despite this silence, believe that this process of change is also registered among the Haldde residents, it is because in 1917 they had a distinguished visitor. No less than Betzy Kjelsberg, the well-know women's activist visited the top of the mountain. She was especially known as being active in connection with the fight for women's right to vote. Betzy writes about the victory for this in her unpublished memoirs.

> At last a complete victory was achieved in 1913. From that point on we attained our power of authority as Norwegian citizens. Some of us sat in the parliamentary box for the public and listened to the negotiations and I cannot describe how happy we were when a final decision was reached for the cause. One of our club members was Gina Krog, who worked with us for many years, and was one of the leading front figures in our campaign. She was just about to travel to an international conference about rights for voting in Hungary. The meeting was to be held in Budapest, and I shall never forget how radiant she was when she said that now she could tell

---

[498] Devik 1971:23.
[499] Devik 1976:19.

them in Budapest that Norwegian women had been given full authorative power, as Norwegian citizens in their country.[500]

Dagny does not write about Betzy Kelsberg's visit in her letters. What is the reason for that? Mrs Kelsberg was a well-known figure in the Norwegian women's liberation campaign and in the fight for women's right to vote that lead to a victory in 1913. It is not at all certain that Dagny was interested in politics. Although her posthumous reputation in the family shows a woman that was involved in the local community. It is quite possible that Dagny left out information about quite a few discussions, concerning political situations in her letters to her mother-in-law. Perhaps her mother-in-law was not particularly interested in political topics? It seems that Dagny kept within the conventions of the contemporary society's expectations of women.

That she as a woman was engaged in what happened around her, is shown through an incident from the Second World War. Dagny chose sides, like many other Norwegians during the war, and hid a fleeing Jewish woman in her flat in Colbjørnsgate in Oslo. The Jewish woman escaped the Germans, but Dagny's role was discovered and she was sent to Grini, the German prison of war camp outside Oslo for a shorter period. Not even that episode is written about in her memoirs. Does this mean that a political engagement was irrelevant to her family situation, and that letters from her life on Haldde were of a more personal nature in which she just concerned herself with domestic issues? Or does it mean that Dagny was controlled by norms and conventions about what one shares with close family members, children and grandchildren? These questions are left open and would probably be given different answers depending on who expresses an opinion concerning them. What is interesting is that nothing of this kind is mentioned in any of the written material she left behind.

That Dagyn's path crossed Betzy Kjelsberg's is a fact. The active campaigner of votes for women and her friend, Johanne Sveen, visited Haldde. Kjelsberg wrote a thank-you letter, formed as a poem, after the visit, in which it is the men who are mentioned most of all. In the letter Dagny and Mrs Devik are mentioned as, "the other half," while the men, "creative gentlemen'" were praised for being such good hosts.

The thank-you letter formed as a poem:

---

500 Printed in periodical "Kvinner sammen", Women United, no. 3.2012:35 Stemmerettskampen. [Suffrage disputes for Votes for women]. www.fokus-kvinner/PageFiles/7648/Kvinner_sammen_3_2012_lav pdf.pdf.

To the Haldde residents!
Now both winter and spring have gone
and next summer and soon the whole year.
Since we crawled up
To the high Haldde top,
Where we viewed the whole surround
And saw many weird things
Both on and under the ground.
Of all the strange things there north
We gladly can remember this:
The dinner for us they made,
You masters' of creation - let cake be served,
We had only wonderful gourmet dishes to taste.
Therefore it is not surprising that
The Haldde top has become dear to us
There we the women, felt our power,
Standing strong in magnetism's pomp and splendour,
When Northern Lights' men can cook
(Otherwise they trifle with the world's axis spinning-wheel
While pondering over the stars, sun and moon's courses,
Yes, dabble a bit in the Lord's plans)
For us they they struggled with pots and pans
And beautifully served plates full to the brim,
We will always remember-with delight-not pain
The cod roe in sauce,
Also delightful apricots
Like a mighty storm wind they captured both our hearts.

Here in the south high prices are colossal
Sugar rations now for everyone
Now too we can hardly get a kilo of anything
Mrs Kjelsberg makes decisions, and works with rationing.
Yes now we remember and long to be up
to Bossekop
Where the Northern-Lights' men + t'other half
Are linked with body and soul
To a 'Sugar top' so firmly
That it will stand, even if the world collapsed.

With very best wishes
Johanne Sveen, Betzy Kjelsberg
25th November 1917

It was in August, before the snow had come, that these women visited Haldde. What was it that brought them there? We do not know. What did they talk about during dinner? We know that Betzy Kjelsberg sat on the central board for the Liberal Party, and had some common political interests, at least with Birkeland, and probably also with those that worked closely with him, Ole and Olaf. We also know that she was the leader of the Norwegian Women's National Council, a governmental state factory inspector and Drammen's first woman cyclist. This lady was both interested in sport and engaged in politics for women's liberation. She established a Domestic Science college for housewives in 1899, which was the first professional college for Home Economics in the country. The poem actually hints that she works with rationing.

Housekeeping matters interested her greatly, and later she became the chairman of the board for the Norwegian Government's Teacher-Training College for Domestic Science teachers. We wonder whether there was any discussion about the Domestic Science school for housewives in Kåfjord when they visited Haldde? In Betzy's memoirs she tells us that her visit to Kåfjord together with Johanne Sveen, was first and foremost to visit the Domestic Science College for housewives, which she had inspected and found that it served its purpose very nicely. It was housed in the previous Copper Work's Director's residence. She had been completely surprised to see, "such a lovely garden and how many lovely flowers they had."[501] Haldde is not mentioned, but it is evident that the two guests carried on to Talvik after the visit, in shopkeeper Heitmann's motorboat.

Betzy Kjelsberg belonged to the middle-class part of the women's liberation movement, but she came from more humble circumstances and was interested in factory workers and their working conditions. What was discussed about the social situation in that region that we can assume was something, which all of them were preoccupied with? It is not inconceivable that comments were made about what was influencing the political debate in Alta at that time.[502] It was especially among the younger generation in the area that socialistic ideas were gaining ground. In the world situation and financial problems, a revolutionary inspired youth association was founded in Upper-Alta.[503] Maybe while they were sitting round the dining table the socialist Kristian Heitmann was mentioned, as he had become the mayor in Alta, and he was a relative of their good friend the shopkeeper in Kåfjord.[504]

---

[501] Note from memoirs recorded in 1950, National Library, Ms 40.2154, Handwriting collection.

[502] Eikeset 1098:184.

[503] Minde 1983:60.

[504] Kristian Heitmann was the Mayor in Alta in 1917 and was deputy for the social democrates 13. May-13 June 1916. Eikeset 1998:180.

Distinguished visit of active campaigner for women's rights, Betzy Kjelsberg, Johanne Sveen and Mrs. Spilling. Archive: Private, Hennige Astrup.

Birkeland's close friend Kaja Geelmuyden was active in the women's liberation milieu, where Betzy Kjeldsberg was in a leading position. The 28th January 1916 Kaja Geelmuyden published an article in the English women's magazine, "Votes for Women". The article was entitled, "Women voters in Norway" and gives us an idea why Dagny says nothing about political topics. In her article, Kaja explains the Norwegian political system, and the growth of women's right to vote, how this vote is used and what arguments have been for and against this. She explains this in a text that covers about three pages:

It was in 1901 that women first received the right to vote, but only in the municipal elections. Even though it was an enormous step forward, many were aware from the very beginning that if they were to have the possibility of representing a new element into public life, they would have to form their own organization and not be integrated into the existing political parties. These were so firmly rooted in interests for men and their point of view that it would be impossible to show clearly the women's point of view, if it deviated from those. Most women had no experience in

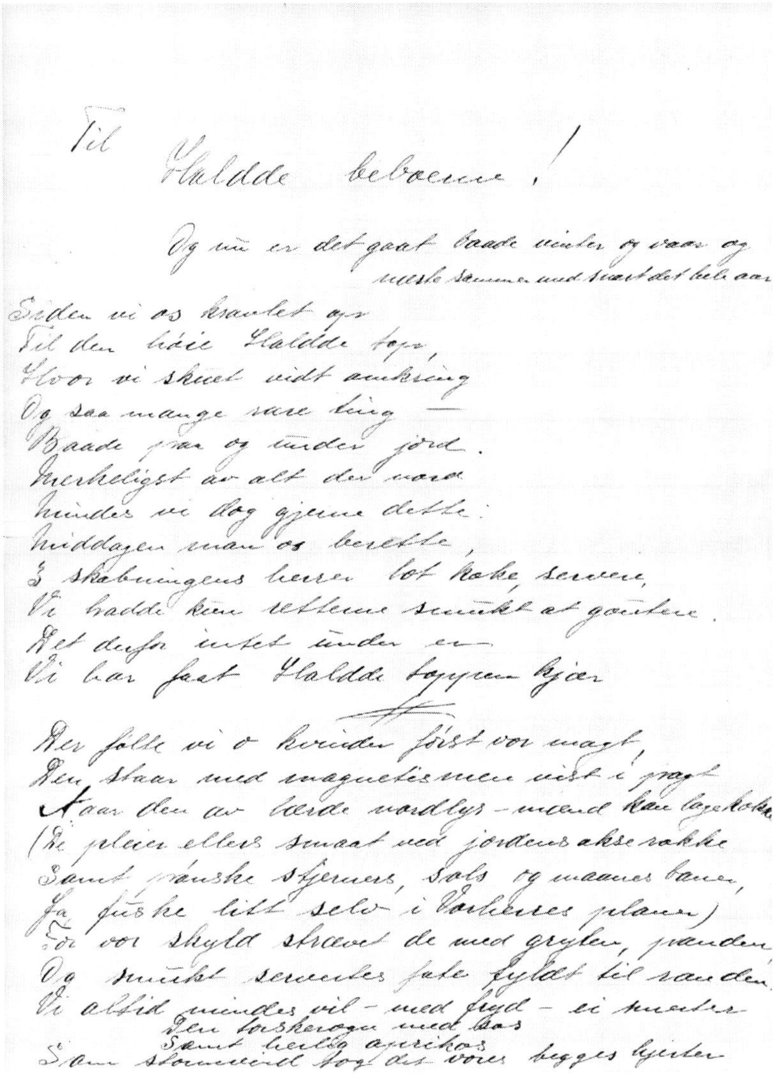
Til Halldde beboerne!

Og nu er det gaat baade vinter og vaar og
mest sammen med snart det hele aar
Siden vi os kravlet op
Til den høie Halldde top,
Hvor vi skuet vidt omkring
Og saa mange rare ting —
Baade paa og under jord.
Merkeligst av alt der vard
Minder vi dog gjerne dette:
Middagen man os berette,
I skabningens herrer lot koke, servere,
Vi hadde kun retterne smukt at goutere.
Det derfor intet under er
Vi bar fast Halldde toppen kjær

Der følte vi o kvinder først vor magt,
Den staar med magnetismen vist i pagt
Naar den av lærde nordlys-mænd kan lage koke
(De pleier ellers smaat ved jordens akse rokke
Samt grunske stjerners, sols og maanes baner,
Ja fuske litt selv i Verdenes planer)
For vor skyld stræved de med grytene, panden
Og smukt serveredes fate fyldt til randen.
Vi altid minde vil — med fryd — ei mindre
Den torskerogn med bos
Samt herlig aprikos
Som stormvind tog dis vores begges kinder

First page of the thank-you letter, formed as a poem, from Johanne Sveen and Betzy Kjeldsberg. Archive: Private, Hennige Astrup.

public matters and just kept to their old habit of believing what they were told by their husbands and brothers, Kaja wrote. The first attempt to organize voting was in a Local Government election, which, "The Union for Women's Suffrage", was responsible for. The result was that two women were voted into the local council in Kristiana. Kaja Geelmuyden writes, "After this the political parties redoubled their efforts to frighten the women back into the party organizations, because if women were to be elected at all, they preferred those who were without any prominent opinions of their own."[505]

That success was restricted, she explains, is because men's organizations are rich, while the women in Norway, "always suffer from the lack of funds; the few that are well-to-do are usually Conservative, and do not take an interest in any social questions, except perhaps charitable ones."

Even worse for women than the lack of money is she claims, that women do not have a journal of their own to write in: "Gina Krog's otherwise very useful paper, "*Nylaende*", is too small, and appears too seldom, only twice a month, to exercise any very wide influence." Despite this, women have nevertheless been able to express their views, because many progressive men like to hear what a woman has to say, "as long as she does not go directly counter to his interests," she writes, and says that in 1907 women succeeded in extended voting to include those who paid taxes or were married to someone who paid taxes, so that finally in 1913, "we received the full political franchise on the same terms as it is granted to men."

There was no strong opposition to this, as Kaja Geelmuyden explains this in her own words:

> According to the opinion of Norwegian men, they have a strong dislike of being regarded as old-fashioned. The long years during which they have worked side by side with women in practical life, has contributed much towards accustoming them to seeing women more or less everywhere.

Nevertheless she still thinks that men cannot be completely satisfied that women are getting more power. She thinks that maybe they have thought quietly to themselves, "Even if women get the vote, it will be a great many years before they learn to use it effectively, and we shall be dead before then!"

The main dissatisfaction among women, she believes, is because they have still not been able to vote even one woman, as a member of the National Assembly. The lack of women in the country's National Assembly is especially unsatisfactory now that they have just had the autumn session, in:

---

[505] Geelmuyden 1916.

> The National Council of Norwegian Women (*Norske kvinners nasjonalråd*), which was held in the "*Storting*", (Parliament). The meeting was popularly called "the Woman's Parliament," and the manner in which Fru Betzy Kjelsberg led the proceedings, aroused the admiration of all.

In her article, Kaja Geelmuyden is presented as being a well-informed feminist, with a clear and critical point of view and courage to present it publicly in a newspaper column. It was Kaja that Birkeland chose to confide himself in, the spring of 1917, concerning the problems he had experienced in Egypt.

Politically dramatic events were occurring in the world at the beginning of the 1900's. Dagny and Ole came to Finnmark and they were to live not very far from the Russian border. Ole had plans of collaborating with Russia. Is it conceivable that the discussion during the dinner party, that Betzy Kjelsberg attended, had Russia as a topic?

On Haldde they subscribed to a number of newspapers, "*Aftenposten*", "*Morgenbladet*", and a number of newspapers from Finnmark, "*Tidens Tegn*", is also mentioned in letters. Occurrences and news about people geographically near were often referred to. In Kirkenes, very close to the Russian border, Ellisif Wessel was very active in political debates.[506] She came as a young doctor's wife to a mining community in Kirkenes with her husband doctor Andreas Wessel. They were shocked to see the extreme poverty that met them. She was a keen newspaper writer.

Those living on Haldde must have been aware of the socialistic ideas that were developing, and knew the people that actively took part in the development of new political opinions. They probably also knew about Ellisif's protests and support to the workers, and the revolutionists. Even so they do not write about it. There is very little about the war in the letters, nothing about the revolution in Russia. Nothing about the Czar´s family being murdered, and they do not write about political murders reported in the newspapers. Nothing about politics is mentioned.

---

[506] See Wikan 2008.

Ellisif Wessel travelled around Finnmark, and took photos of people working. Archive: South-Varanger Museum/ Grenseland Museum

We know that Birkeland was politically engaged, as this has been made quite clear in connection with his involvement in the dissolving of the union with Sweden, an involvement he shared with, for example, Sem Sæland. His close friend, Professor Amund T. Helland, was an acknowledged socialist, who wrote critical contributions in the most left-wing sympathetic newspaper of that time, "*Morgenbladet*". He and Birkeland often voted for the same things in faculty meetings. Helland was, like Kaja Geelmuyden, one of the few friends with whom Birkeland discussed personal problems and issues.[507] It is probable that his students, as young intellectual academics, shared his political views. Such information does not appear in any of our main source material.[508]

There were similarities in the daily lives of the Wessel and the Krogness couples, at the beginning of their stay in Finnmark. They bought a reindeer to use as transport; they employed young girls and boys from nearby. While the girls were maidservants and helped in the house, the boys took part in the transport of wares and driving the reindeer. Both couples mixed privately and socially with the clergyman, shopkeeper and other bourgeois people living in the rural area they now lived in. Although while Ellisif was in the middle of everything that was happening as the wife of a district doctor, Dagny was more isolated in her life as a housewife on the top of a mountain with her research husband. Ellisif

[507] Egeland & Burke 2005:161.

[508] Both Amund Helland and Kaja Gelmuyden died in 1918, a year after their close friend Kristian Birkeland. See Egeland & Burke 2005:161.

experienced poverty, illness, poor living conditions and injustices in the community and reacted to it. Dagny experienced this to a lesser degree and definitely did not write about any possible reaction.

The only time we see any sign of a class perspective is when Dagny comments that Hoel is going to marry, and the Lukkassen family have moved up to the top of the mountain. How should the flats be divided? Dagny thinks that the Lukkassen family should have the largest flat, as they are a family and need more room. But Hoel is the 1st assistant, and therefore has a right to the best flat. "There we have the world's injustices again," concludes Dagny.[509] Later in her life we know that she came the nearest to rebellion when attending a concert during the Second World War, when the names of Nazi sympathizers were called out from the stage.[510]

Both Dagny and Ellisif were middle-class citizens born into a family of doctors, and had lived on Dovre, but there was twenty years between them, Ellisif was the eldest. Dagny experienced that all her brothers were given a higher education, but not her, and Ellisif attended the newly established middle school for girls, Nissen School in Kristiania, in 1882 before she married. The couple's home in Kirkenes was full of books. Ellisif read a lot. On Haldde the library was inadequate.

Women working with "senne" grass. Photo: Ellisif Wessel Archive. South-Varager museum/ Grenseland Museum

The two women with a similar cultural and class background had different social positions through their marriages. The young Ellisif read Marx and Lenin, learnt Russian, was a revolutionist and stood on the barricades, an individualist, exposed to hard criticism. She had contact with Lenin and Trotskij, and was called an interesting person by

[509] Letter from Dagny to her mother-in-law 13.11.1914.
[510] Letter from Dagny about what happened. This belongs to the family.

Alexandra Kollontaj.[511] The young Dagny sat safely on the top of a mountain looked after her children and wrote reports to her mother-in-law, while her husband was travelling on important research errands. These two women had completely different daily lives and very different experiences with pregnancy, birth and children. Ellisif aborted a number of times. She gave birth to one child, who died. After a long period in great sorrow and depression, she began to engage herself in social conditions and politics. There was no need for her to use any of her time looking after her own children. Life for Dagny was very different. It was the life of a family with small children and full of everyday events.

When the Haldde families moved down from the mountain, it was to continue with the work they had begun. They had created their own workplace, the Geophysical Institute; in a somewhat less isolated position. Also here, Dagny's everyday life, where things happen around her, is what she tells her mother-in-law. At the same time even here their social life becomes more out-going, and their circle of acquaintances increased through Ole's work and cultural engagements.

The work with the further development of the institute in connection with the museum also included contact with the local municipality's political authorities. Therefore it is possible that they came into contact with another important woman pioneer from the north, Gitta Jønsson (1869–1950). Dagny must have known her as both of them lived in Tromsø at the same time, and Gitta Jønsson was prominent in public life. She became the first woman to be a member of the executive committee of the town's local council from 1913, where she made her position clear by launching sexual information and counselling on prevention for women. This was strongly opposed in the male-dominated and middle class executive committee. She was a member of the school council's poverty committee and a factory inspector. She probably knew the factory inspector Betzy Kjelsberg.

Gitta Jønsson led "*Tromsø Arbeiderkvinneforening*", (Working Women's Association), for 35 years. The association fought for women's rights and contributed in the fight for a general right to vote for women in 1913. This woman pioneer, Gitta Jønsson, came from relatively poor circumstances in Tromsø. Her father was a fisherman and later an Arctic Sea skipper. There were nine children in the family, and Gitta was the eldest. There was little schooling. She travelled to Kristiania and took a domestic position, and established an acquaintance with a growing working

[511] Wikan 2008:187.

class, experienced enormous class differences and open exploitation of different kinds of treatment, mainly directed at the working class.[512]

This provided the grounds for the ideas she took with her when she moved back to her hometown, Tromsø, together with her Swedish husband. There she engaged herself with women and the lower classes. Her experiences from when she was growing up had shown her how poverty and continually giving birth weakened women. The situation was especially bad for those who gave birth outside marriage. In 1916 with Katti Anker Møller, another women's liberation fighter, she took the initiative to start a national information and childbirth campaign, which covered both birth control and hygiene.[513] On the 2nd of May 1919, the year after Dagny and Ole arrived in town, "*Tjenestepikers forening*", the "Working Women's Association, for women domestic servants", was founded in Tromsø. The initiative for this came from the "Labour Party's Women's Association", where Gitta Jønsson was an active central figure.[514]

This was a part of life that Dagny knew well from her own everyday life. Yet we cannot find any reference in her letters to her mother-in-law, about the working class and their politically oriented women pioneers. A wider knowledge and information about the women's movement's work, feminist ideas and women's emancipation, first benefited this part of the country, paradoxically enough, through the establishment of the university, and the initiative of the better off Haldde researchers.

In her memoirs Dagny describes herself as a child on Dovre:

> I was quite lonely as a child, as one could not talk about everything to one's parents. I sneaked a look in Uncle Gustav's anatomy and medical books that belonged to my father, but they were in German, so I just looked at the pictures. Otherwise I ran a lot in the woods and fields alone, decorated myself with garlands of twinflowers and dreamed lots of strange things.[515]

In the first letters from Haldde a longing to have someone to talk to is repeatedly mentioned, "[…] because you know, loneliness creates so many thoughts and questions […]", the young Dagny writes to her mother-in-law and one suspects there is quite a lot she worries about.

---

[512] Grønbech, 2012:3–8.
[513] Grønbech, op. cit.
[514] Nielsen 2002:105.
[515] Memoirs written by Dagny for her grandchildren, 1976.

But throughout the years Dagny adjusts herself to her role as a privileged housewife, as the "other-half" of a husband in a prominent position. Something had changed inside her.

# CHAPTER TWENTY FOUR

# FURTHER WORK

A selection of Norway's leading researchers gathered together at a Geophysical Congress in 1919. Both Olaf Devik (back left) Ole Andreas Krogness (Number 3 from the right standing) and Hilding Köhler (furthest out to the right) are among the participants. Carl Størmer to the left in the front. Archive: The World Heritage Rock-Art Centre, Alta Museum

*Yes they were happy days in Tromsø. In 1928 we moved to Bergen, your grandfather was Professor at Bergen Museum. It felt like the Middle Ages in the beginning. Tromsø was a city full of life during those years. It was called the Paris of the North, but in Bergen everything seemed old and the people old-fashioned.*[516]

[516] Memoirs for grandchildren written by Dagny, 1976.

The Institute in Tromsø was able to take care of the weather-forecasting service much better than the Haldde observatory. On the other hand Haldde was still an important place, in connection with the research of meteorology, as experienced by Köhler, who from 1918 was the new Director there. He held the position until 1926 when Haldde observatory was closed down. Ole was the Director in Tromsø in addition to having the responsibility for the department of terrestrial magnetism and the Northern Lights. Devik had the main responsibility for weather-forecasting services in Northern Norway. Devik left Tromsø in 1922 to take up a post as senior lecturer in physics at Norway's Technical College in Trondheim.[517]

Margit Elisabeth Köhler with her sons Max Leonard and James Rurik, together with Mrs Lukkassen and the maidservant Gudrun Dagenborg from Kåfjord. The photo is taken in the "*Borgen*" (the Castle) in Kåfjord. Photo: Hilding Köhler. Archive: The World Heritage Rock-Art Centre, Alta Museum

Tromsø really wanted to have their own department for Northern Lights' research, and in 1925 they managed, in cooperation with Professor Lars Vegard, to get finance from the Rockefeller Foundation, as well as a

[517] See his further career in Devik 1971.

Norwegian State grant. As a continuation of the observatory on Haldde in Finnmark, the Northern Lights' Observatory in Tromsø officially opened in 1930.[518] Devik and Krogness, also further developed the meteorological station network in Northern Norway, through their work at the Geophysical Institute. The Station Network consisted of 45 meteorological polar stations and a magnetic station on Dovre. Smaller Arctic stations were built on Spitzbergen, Quade Hok in 1920, Jan Mayen in 1921, East Greenland and Bjørnøya.[519]

From the Geophysical Institute Ole administrated the Polar Stations, and was responsible for expeditions in 1918–1928. As a scientist he engaged himself in giving lectures about his work, and in addition to this he taught at the sixth-form college. Later it was said, "He was not a gifted speaker, and he could have difficulties due to not being well enough prepared, but he coped, and it was always interesting to hear him.[520]

In addition to research, the Geophysical Institute in Tromsø had some practical tasks. One of them was collecting data from different weather stations in Norway, so that the best and correct weather forecasting could be given. As no special organisation was responsible for this important issue in the Arctic region, Ole took responsibility for it. He was the one who arranged rescue operations for hunters and fishermen, when they got trapped in the ice or shipwrecked. If one of the stations had problems, he found a way out, as for example when a participant of one of the expeditions to Greenland became ill because of scurvy. Devik relates, "Krogness started reading about vitamins, quite a new thing at that time, and got the others on board to let the peas sprout, and to give these to the sick person to eat. And the man was saved."[521] He also assisted scientific expeditions either at the planning stage or when they were on their way out to make new discoveries. Devik describes it thus:

> There is quite a bit of risk for those who sail in the Arctic, and on a number of occasions expeditions requiring help became necessary. Not everybody understands how much careful planning and work is done beforehand, but the person who is leader of the institute in Tromsø is aware of it![522]

[518] Devik 1976:218.
[519] Devold 1940:11, Drivenes & Jølle 2004:220.
[520] Evjen 1941:10.
[521] Devik 1976:114.
[522] Devik1935:165.

The Lukkassen family moved to Tromsø and Karl was the caretaker at the Geophysical Institute in 1925. Archive: Private: Ann Marie Sørensen

On top of his normal job Ole also worked with the establishing of the weather service stations on Spitzbergen, Jan Mayen, Bjørnøya and Greenland. In connection with this work he was also involved in the conflict about who had the rights on East Greenland; Denmark or Norway? That a weather service was already established was for the benefit of those working in the Arctic region. To be able to utilize the weather services more efficiently than just using a telephone, he had sent a radio station to East Greenland. As far as he could see it was natural for huntsmen, whalers and fishermen to also work on land. The Danes were not in agreement about this, and disagreements broke out as soon as the weather services began to be sent over the radio. The otherwise mild Ole Krogness was angry. He had arranged for a station to be placed in an unknown area, something that from his point of view was an advantage for the whole of Europe's weather services. He felt they should have been grateful.

> It is so exsasperating! Here our boys with hard work, hardships and daring have found a hunting area, which they alone are able to use, and where

> others have no interest, and now they may risk being driven away. It is enough that the Danes managed to swindle the land for themselves earlier. Now they want the east coast as well, as soon as we have shown that something can be done there.[523]

The argument concluded with the East Greenland Agreement in 1924. Now weather-forecasting services could be spread more effectively. The establishing of a radio sender increased the public's interest for the radio, and led to Tromsø having its own broadcasting service already in 1925.[524] On the south of Tromsøya where the Broadcasting station is built, the road is named after Ole, and called "Krognessveien".

The Hunters hut on Greenland. Archive: Tromsø Museum-University Museum

After the experience from East Greenland Ole was part of the annexation of Jan Mayen. Part of the island was already annexed on behalf of the Meteorological Institute. Now in 1921 Ole suggests that they should annex the whole island, as there were no "foreigners" that could declare their interest.[525] Therefore nobody could say they were standing on their toes. Consequently, Ole sent a telegram to the Norwegian authorities with the suggestion of annexing Jan Mayen. The Norwegian authorities were not especially positive, and wondered what they were supposed to do with an island that had no inhabitants, had dangerous waters, and only

---

[523] Devold 1940:11.
[524] Devik1976:115
[525] Devold1940:43.

Norwegians utilized it? After a period of time Jan Mayen was annexed, and from 1929 was made a part of Norway.

The meteorological service for Northern Norway was important in a number of ways. When Roald Amundsen was to fly to the North Pole, Kings Bay, Spitzbergen was chosen as the starting point. A special weather service was established there. Amundsen, together with the five who were to be on the expedition, already arrived there in April to wait for good weather. Meteorologist, Jack Bjerknes, who was on duty, first reported on 21st May that the weather was favourable and Amundsen with all of the members of the expedition left. The plane failed to appear for a number of weeks, and many were afraid that something had gone wrong. What they did not know was that the plane had to land in an ice hole because of motor failure. It was a small plane, but in some way or another all six managed to get back in and fly the whole way back.[526] The Arctic researcher Devold, who has written about many of the incidents from that time in his book, "*Polarliv*", tells about how Krogness enthusiastically came running from his office and burst out that he knew they had survived. "Roald has come back, all six in a machine! Where is the caretaker? We must raise the flag!"[527]

Two years later Amundsen was, together with others, looking for the Italian Nobile's air expedition, which had been wrecked. He never came home again. The 18th June 1928 he flew with Latham from Tromsø and crashed soon afterwards, most probably near to Bjørnøya. Krogness was responsible for the weather service and he took it very hard when they did not come back.[528] Dagny comments, "They were not very interested in weather services, just rang and said: Yes, weather forecasting would be appropriate now."[529]

Simultaneously with his work Ole was active in Tromsø's cultural life. As the city was to become a scientific centre of growth, culture was a natural part of this development. Ole engaged himself in work for Tromsø Museum, Polytechnic Association and Tromsø's broadcasting programmes, but he was especially active in the city's musical life, where he played a central role.[530] There was always music in Dagny's and Ole's home, and whenever it was possible he arranged a quartet in which he played one of the instruments. Dagny and the children were also encouraged to take part

[526] https://snl.no/Roald_Amundsen
[527] Devold 1949:41.
[528] Memoirs for grandchildren written by Dagny, 1976.
[529] Memoirs for grandchildren written by Dagny, 1976.
[530] Devik 1976:115.

in musical activities. Synnøve became a competent pianist during their time in Tromsø, and often played the piano with her father.

Latham just before the beginning of Roald Amundsen's expedition to find Nobile. Amundsen is sitting on the back of the plane. Archive: Tromsø Museum-University Museum

It is said that if Ole did not have a musical instrument that he needed for a group of musicians, he went down to the Coastal Liner quay to wait for a boat to come in, and asked the arrivals if any of them played the instrument that the group lacked to make it complete.

Ole took every opportunity to play. One of the meteorologists later described how he himself was dragged in to play his violin for musical events:

> On the way to Haldde Observatory I was going up to visit him, while the boat was docked in the quay. "Yes. You can have dinner with us", Krogness said. "Here you have a violin, and then I will play the cello and my wife the piano, while the maid makes dinner. We can play Ave Maria." No protests helped. However, I was just about to leave, when Krogness suddenly thought that I could take some letters with me to Director Köhler. "It will not take very long to get them finished." As I waited I became more and more nervous about the Coastal Liner, after a hurried sprint to the quay, only a 100 metres to the boat, it steamed northwards with my

entire luggage. Well Krogness kindly invited me to stay with them and take the next boat and to continue practicing Ave Maria.[531]

Sometimes Ole had to go down to the quay to ask if any of the passengers played the instrument he needed. Photo: Thoralf Øien. Archive: Perspektive Museum, Tromsø

It was also said that when new assistants were to be appointed, those who were preferred were those who could play an instrument, if there were a number of applicants to choose between. But also other musically interested people in town contributed to Ole's concert activities. Dagny tells us in her memoirs that Doctor Dybwad-Holmboe, always chose an assistant with a passion for music when he was looking for a new doctor. Then he asked Ole:

> "What do you want now, Krogness?" Then he would get someone who played the instrument they lacked. There was Doctor Erling Hiorth, violin, a Doctor Hall with his cello, among other instruments, and Hans Jacob Ustvedt who sang. We had a choir once that performed The Elf King's Daughter by Niels Wilhelm Gade [...]. A lot of music was played and many artists came to Tromsø during the years we lived there. After a concert they would often come to us for coffee and cakes, and sometimes

[531] Evjen 1945:10.

> there were formal social evenings at other people's homes. They were happy years in Tromsø.[532]

In a way music and a cultural life was a part of the scientific activities that took place, and which were part of the process for further development. Dagny remembers her life in Tromsø as being a lovely time:

> It was said that people had a lot of money after World War 1 and parties were held all the time. During the Polar winter months it was best to be in motion, dance and enjoy oneself. Those who lived a conventional and calmer life often suffered from a lack of sleep. Children could also have problems sleeping. I often sat in the evening and tried to get them to sleep by singing for them, even for the older school children.

Researchers tried to cure insomnia by:

> […] making a "*Høyfjellssol*", (sunlamp), an arc light between electrodes, It gave a strong light (ultraviolet rays) that could give us a tan when we sat in front of it. We would take Oa and Synnøve upstairs in the evening and let them sit in front of it for 20 minutes and then they would fall asleep. When we had parties we used it as entertainment, and had a ladies and gentlemen's show. It was also introduced into schools, where children would receive light treatment and cod liver oil during the Polar night period.[533]

Apart from the music, and the parties and the life of research there was also a daily life where Dagny reigned, and where she and the maids always had their respective tasks. These tasks were in different ways involved in and part of the scientific and social milieu, which became the start of the University in Tromsø.

Those who spent the winter on Jan Mayen, for example, used to have pigs with them. One year they bought too many and Ole suggested that he could take one of them for his family. The pig was given the name Karoline, and was from the beginning thought of as a contribution to the housekeeping:

> We were a large family now and needed a lot of food; we were eight with the maid. Yes, we got the caretaker Johansen to make a pigsty from a piano box. He drilled air holes in the walls. I asked him if there would be enough air, and he said he was sure there was. The pig grew and thrived the whole of the autumn. Then one Sunday morning the maid, Konstanse,

---

532 Memoirs for grandchildren written by Dagny, 1976.

533 Memoir s for grandchildren written by Dagny, 1976.

> came running into the bedroom crying, and threw herself by my bed and cried out, " Oh Mrs Krogness, Karoline is dead!"

Upon which Dagny sat up in bed, thought quickly and said, "We must bleed it so that all the blood runs out!" The pig had been fine the previous day and had been outside for a walk along the road in the snow. Dagny got up and found a long kitchen knife and ran outside:

> On the way Engineer Wilhelm Lund came past in his Sunday clothes carrying a psalm book on his way to church. I shouted to him, "Can you bleed a pig?" He came with me to the piano box and there was Karoline, poor thing, in the door opening, she had probably been suffocated.

Before they start using the knife Ole came. He was calmer than Dagny, and arranged for proper treatment of the dead Karoline. Dagny continues to tell about what happened:

> We sent the intestines to the vet, and he said that we should not sell the meat, but could eat it ourselves. I usually tell people that I cooked the chops and let my husband and children eat them first [...]. There was a lovely Christmas ham. The Christmas guests had meat that was bought, but we ate our pig.

Based on everyday small dramas, music and cultural arrangements and the scientific, practical and administrative work at the Geophysical Institute, an idea began to sprout that led to the establishment of the Northern Lights Observatory in Tromsø, an event, which pointed forward in history. As an extension of the activities carried out at Haldde observatory in Finnmark the foundations were also laid for visions, both for Tromsø, and for the region's future development.

Dagny, Ole and the children left Tromsø in 1928, as Ole was to take up the Professorship in terrestrial magnetism and physical cosmology, at Bergen Museum. In the application for the position Ole pointed out that the practical organizational work involved in the construction of the weather forecast service was very complicated, and had taken up most of his time and working capacity throughout many years. Therefore there had been little time for scientific work, and especially work on material about the Northern Lights and terrestrial magnetism.[534]

Ole got the position in Bergen and when it was known that he and his family were going to move, a big party was arranged as a thank-you for everything that he had done. There were three hundred guests at the party,

[534] Application dated 20.05.1928.

The white mountain plateau was exchanged for the city streets in Tromsø. Photo: Boye Strøm. Archive: Sparebank 1 in Northern Norway, Tromsø.

It is possible that it looked like this when a banquet was arranged for Dagny and Ole in Tromsø Sparebank's Assembly Hall. The occasion here is unknown. Archive: Perspectivet Museum. Tromsø.

it was held on a Saturday evening in the, "*Sparebank's*" Assembly Hall. The whole city was represented. There were many speeches, a lot of music and toasts, and dancing through the night. Only Ole participated. Dagny had already gone to Bergen. The party guests sent her a greetings telegram of three hundred words, with everyone's signature.

The eventful evening was properly reported in the newspaper.[535] We find out that they ate through a menu of roast reindeer and cloudberry compote, to the tones of the cinema orchestra, until the former Cabinet Minister, Rye Holboe held the ceremonial speech, and among other things had this to say:

> Dear Krogness!
> We have come together to pay homage and to thank you. The invitation to the party is from the City Municipality […]
>
> It is Tromsø city that celebrates you, because the city owes you a thank-you. I have been asked to interpret the feelings that inspire us all. I do that willingly. It is an easy task. […]
>
> You came here after a number of years of scientific work at Haldde observatory, and looked for more favourable conditions for your work. It was not difficult for you to persuade us. You have a fortunate ability to communicate some of your belief and what you are ardent about. […]
>
> Thanks to your great energy and competence, and your, and believe me when I say, enormous distain for difficulties, whether they are to do with the Government, a Parliamentary committee or the State audit, you got your plans realized.
>
> You were soon popular in this city. We do not understand so much of the scientific work. But we understand that a Geophysical Institute was a good thing.
>
> The next phase was the weather forecasting service. You yourself think that you must share the honour with others, but without your energy and interest, the weather-forecasting service here in the north would never have got started […]
>
> We all understand the benefit of that work, and especially those who travel out to sea.

---

[535] This was printed in a side column of a newspaper the family have. The name of the newspaper was not kept, but it was a newspaper from Tromsø.

> The third phase was the Northern Lights' Observatory. Even though we may not quite understand that work, all of us are aware that it has a great value […]
>
> In Tromsø we have a great and deferential respect and interest for scientific activity, which is far beyond our comprehension, and we know that you still have ample plans for the future […]
>
> What is quite certain is that the work done to develop the cultural centre here in Tromsø into a cultural centre in Norway, during the last couple of generations, at least, no single person has given so much work, so much support and is deserving of so much merit than you.

In his thank-you speech Ole said:

> I stand here with a strange feeling. Such a festive evening as this should be for men who have done something. What I have done is far too small for such celebrations! The valuable work that has been carried out here is not deserving of just a single person getting the merit for it. There are a number of factors that have worked together to achieve such a result.

He gave homage to Birkeland, Bjerknes, Devik, the Chemist Dahle and other important men, Tromsø city and Tromsøysund Municipality, hunters and fishermen at sea, expeditions and pioneers in Arctic regions. He talked about his happy years in the north, which he will never forget and said, "I wish all the best for Tromsø, and that it will eventually continue to grow into an even more important cultural centre for our country."

For Krogness after he left Tromsø in 1928 from
The Weather Forecasting Service for Northern Norway and Tromsø.

Ode
To Krogness
Or to write the above valedictory
OD

OK
I
1) In the beginning he sat in Egypt
and painted and wrote numbers in cryptography

2) It was so confoundedly hot there in the sun
so Krogness he left, and up to the Pole he did run

3) "Upwards" is the motto for researchers being cold
and Krogness up to Haldde he climbed, we are told.

4) There he stood and photographed Corona.
The night was rather long, especially for his wife.

5) The poet writes odes
Krogness discovers weather periods

6) Thoughts teemed and create at last
the earth's most northern institute.

7) So one ravages the coast with old Zoe
to show the world that work is done.

8) It is not only with, "Armauer Hansen",
that scientific men, in this country, perform dances.

9) He measures magnetism everywhere,
down into valleys, and up over heaths.

10) And Zoe, who wants to come up in the country,
must fish below in waters, for terrestrial magnetism

11) The work load grows and finally creates,
our absolutely most northern institute.

II
So over the mountains to Bergen he moved
While Evjen with wife, born Berge, were left behind
They saw him in the direction of the ferry disappear,
and shook and shouted, "Come back soon, my dear."

III
1) And the city lay in hibernation and did not say so much,
from the chains of boredom, when Krogness left.
But the telephone chimed, so that was pretty sneaky
One managed to answer, as that chiming was so beastly.

2) But now he is back and we hope he is man enough
not to go astray for suggesting a third institute up here one day.
And thereby we come to the end, but it sounds, as you know:
That Krogness can stand as Godfather to a university.

# CHAPTER TWENTY FIVE

# LIFE MOVES ON

Ole Andreas Krogness surrounded by his children, from the left, Eva, Oa, Inger, Karl and Synnøve. Archive: Private, Ingrid Løvlie.

*" I was unfortunately left on my own in the thirties, with five children."*[536]

On the 28th May 1934 Dagny's world collapses around her when she loses her life-long partner of 22 years, Ole Andreas Krogness, who died after a period of illness. A far too early a death, many said, a loss for a family, friends, colleagues and the local society, but also a loss for science and for scientific work.[537] Dagny is left on her own with five children: Ole Andreas, Synnøve, Inger, Eva and Karl. Their lives change dramatically; from being part of a vibrant research and musical milieu with

[536] Memoirs for grandchildren, written by Dagny, 1976.
[537] Mentioned among others in Evjen 1945 and Devik 1971.

maidservants and housekeeper, to an everyday with poor economy and little help in the house.

Quite a few books are written about Birkeland, and Olaf Devik has written about his time on Haldde, but Ole Andreas Krogness died before he or others managed to write extensively about his work. As Ole's contribution was so significant in everything that happened on Haldde and its continuation, we will delve further into parts of his life and work which this book so far has only touched upon or not covered.

It is seldom that a scientific researcher has such a good posthumous reputation about what he has meant for people, and about the warmth and generosity he showed to all around him. Among people in Kåfjord today there are still stories about how Krogness was a man who talked to everybody, had time for everyone and took people seriously.[538] Olaf Devik, his good friend of many years, described him as being more of an inspiration than many, amiable and a person who quickly found new friends. Meteorologist Sigurd Evjen, an earlier assistant to Köhler gave this picture of him: "[…] what made most impression on me was his warm heartedness. It was a source of warmth that always bubbled over […]."[539]

Ole was emphatically social; something, which his daughter Eva tells could be an obstacle at times, for example when going skiing with her father. Once they had set off, he always met someone he either knew or did not know, who he had to have a chat with. In the end the rest of the group left him, only to find him eagerly talking on their way back not far from where they had left him.

His social demeanour also characterized his work. Evjen writes that some people thought that Ole Krogness possibly managed to do so much in such a short time because he was ruthless, scheming and ambitious for himself. According to Evjen that was a complete misapprehension:

538 Werner Daniloff, interview 14.07.2008.

539 Evjen 1945:11.

When the family went on holiday they travelled from place to place. They tried to stay somewhere that was as cheap as possible either with people they knew or they rented a room. Archive: Private: Mette Bunting

> What characterized Krogness more than anything was his amiability, a winning disposition, and his ability to make friends. When I met him for the first time, while the coastal liner was docked in Svolvær, he began his enquiry in such a way that I thought. “Good heavens, I would like to take part in that,” and I think he awoke the same kind of thoughts in most of the people he tried to rouse an interest in for his plans. He got laymen to think that science was fun and not just dry numbers.[540]

To create enthusiasm was something Krogness was good at. “Ideas took root and grew around him.”[541] On Haldde he had to use all his ability for science, ingenuity and not least people. He had many ideas and a great capacity for work, always engaged, but with the ability to fully unwind preferably with music. Devik compared Krogness’ talents with Birkeland’s. He had a purposeful toughness that others were not always aware of, and an imaginative ability for improvisation.

[540] Evjen 1945:9.
[541] Devik1976:14.

The Krogness' family car did not have room for everybody so someone always had to stand on the running board when they drove. Archive: Private: Ingrid Løvlie

Despite having both a strong and weak side he was unconditionally charming, and he made cooperation with others a rich and significant experience. Evjen stresses that the variety of tasks and projects did not mean that he was:

> an impulsive person that constantly threw himself out in new undertakings and by so doing forgot the old ones. He always came back to the main aims in his work and really worked, tenaciously towards his goal. Not everything succeeded, e.g. to create a University in Tromsø, but that is an aim that demands generations of hard toil: yet at some point the first seed must be sown.[542]

That Krogness could be challenging to work with cannot be ignored. Evjen says the same:

> Often when I thought that something was finally ready, a third man came into the office, and so Krogness began, "We sit here talking about……" With that, time went which was not always a good thing for a recorded

[542] Evjen 1945:10.

weather warning that was to be synchronized with the clock. He could keep on talking about what interested him until urgent matters could no longer be postponed, and had to be done at the last minute.[543]

According to Devold, Krogness was known for his good and kind heart, he was an exuberant man and, "to a great degree loved by his subordinates".[544] When some of the crew for the Polar stations began to argue or came with outbursts, it was him that was able to meet them with good-natured indulgence. Good training from six years on Haldde is sure to have helped. He knew what it was like to be isolated for a longer period of time, and could calm people down. How popular he was became clear when two huntsmen, Møkleby and Simonsen died of scurvy and tuberculosis. They had dug a hole in the snow to survive in Kobberbukta on Svalbard, but when a boat came looking for them, they were asleep in the hole and first caught sight of it as it disappeared out of the bay, too late to see them. They wrote diaries during the last days they lived, and when Nøkleby wrote his last greetings, it was, "with thanks to Krogness for letting him get to know such a generous person".[545]

Krogness was active in organisation life parallel with his own research. He was one of the founders of the "*Norsk Geofysisk Forening*", (Norwegian Geophysical Commission) Professor Birkeland's Fond for geophysical research and the journal, "Geophysics Publications". Locally he was the initiator of, "*Komiteen for de Vitenskapelige Institusjoner i Tromsø*" (The Committee for the Scientific Institutions in Tromsø) called, "*The Comet*". That committee had as their leading issue a demand for its own university for Northern Norway, and they established a fond. Additionally he was also a member of the committee for the "*Polytechnic Association*", as well as, "*Tromsø Music Association*", member of "*Tromsø Museum's Board of Directors*", from 1918 to 1928, and Chairman from 1921. In 1928 he was elected as honorary member of Tromsø Museum, on his departure to Bergen. Krogness was otherwise one of the founders of "*Tromsø Broadcasting*. "He was a leading force in all cultural life in his locality. He was also the driving force in the city's music life and was an excellent violinist, cellist and pianist."[546]

[543] Evjen 1949:9.
[544] Devold 1940:11.
[545] Devold 1949:6.
[546] Evjen 1945:10.

Right until the end, research was important for Ole Andreas Krogness. Here he is doing one of his countless measurements. Archive: Mette Bunting

He participated in a number of scientific meetings abroad and was chosen in Paris in 1919, to be a member of the, "*Polar Commission*" and the "*Kommisjonen for jordmagnetisme og luftelectrisitet*", (Commission for Terrestrial Magnetism and Atmospheric Electricity). Ole Andreas Krogness was also made, Knight of the French Honorary Legion and Officer of the Italian Crown Order.[547]

Even though Krogness published quite a bit throughout his short life, he was also criticized for not having published more. He was an inspiring source of ideas, and Theodor Hesselberg, the Director for the Norwegian Meteorological Institute, encouraged him to write more about these, and publish them. It is possible that if he had not been so engaged with so much, both scientifically and culturally, he would have managed to write, and create an even greater name for himself, as many people had expected. That would have meant that other parts of his life would have had to be renounced, and that posthumously he would perhaps not have been presented here in the same way.

[547] Devik 1971. *News Paper:* "Bergen Aftenblad" 29.05.1934.

Two serious parents with their five descendants. Archive: National Library of Norway - Photo collection.

Krogness became ill early in 1934. He developed kidney stones, and at that point of time treatment methods were lacking, and nothing like those of today. The illness was a great strain on his body, but even so an illness that both he and others thought he would get over and return to full health. But this developed into uremia, earlier known as renal failure, when the whole body is poisoned and the kidneys no longer function. Throughout the whole of his illness he was constantly on the phone, or welcomed visitors so that he could disclose his latest ideas or ensure that work was being carried out. The eldest children lived at home, while the two youngest, Eva and Karl, were placed with people they knew. Eva tells us that when she came home to say goodbye to her father, neither she nor her brother knew how bad things were. Devik describes their parting like this:

> The next day, when once again, a number of us went to visit him, he collected all his children round him and talked to them in such a way, that they would always be sure to remember. He mentioned, "the new family", who he jokingly called the four who should now form a new council, and embraced all of them. Not just literally, but also with his brilliant glance and his good smile. All of them gathered together in the Lord's Prayer and a psalm, which he himself led. A free and effortless peace and joy that in a strange way lay over it all.[548]

[548] Devik wrote to his children about their father's death. Private possession.

Later Eva said that she hid upstairs, and could hear how her father was in pain when he thought no one else was there.

Olaf Devik wrote down the events that surrounded Ole's deathbed, so that his children should know what happened. He tells us that Ole collected his friends round him and was honest with them, and said that he did not think he would live. Even so they drank to his health, "as if no shadows were allowed to draw nearer". Olaf Devik spent a lot of time by his friend's deathbed and describes it as something very special; there was no feeling of a long drawn out parting, rather the contrary. He writes, "Krogness could also say that he felt completely happy, and he expressed it so paradoxically when he said: It is wonderful to live, but even more wonderful to die."

Kristian Schjelderup,[549] a colleague of Devik's, was one of the many who visited Ole when he was ill. The speech below was published in the local newspaper, Nordlys 30.06.1934, and Dagny sent it to Sigurd Evjen, among others, as he was unable to attend the funeral.

Nordlys 30.06.1934
*By Kristian Schjelderup*

On Ole Krogness' death.
It is impossible to do justice and express in words what we have felt during these last days. There is a strange mixture of sorrow, and an almost festive mood. Sorrow, deep sorrow because we have lost such an indispensable person, but even so with a sprinkling of festivity because we have experienced the triumph of the spirit.

Personally it has been deeply moving, and I feel the need to say thank-you for having been able to be a part of it. It has made a lifelong impression.

We, who daily go round and plague ourselves, and others with petty worries, with quarrelling and clashes and fuss over small trifles, we who speculate and theorize over life and death, and are full of uncertainty. Yes exactly, doubt and fear, we have in some unforgettable days stood face to face with a person, whose

[549] He was known as a society personality, a disputed radical and critical theologian in the the Norwegian church. Schjelderup was concerned with science, and was from 1930 a religeous psychological researcher with Christian Michelsen's institute for science and intellectual freedom. He cooperated with Nansen about the creation of Nansen School. Bloch-Hoell 2009 and Repstad 2009.

wonderful display on his deathbed has been an anchor for us in our own fight for life. We who are unsure and fumble and often fail, we have been witnesses to a spiritual authority which in all its simplicity has carried with it, eternity's security. We who would really like to believe, but often find it difficult, we have for a moment seen into brilliant, clean, explanatory eyes, that reflect another world's grandeur. Round us we noted some of that soothing peace, in which the divine soul can be found. And therefore have in some way been more certain, it is not just sorrow and loss we now experience, but also pleasure and hope. I personally have never stood at a deathbed that has felt so forgiving. Despite the meaninglessness, when we think about him personally, his home, his work, it nevertheless has for me given some kind of meaning and by giving others of us faith in the return of the spirit.

There is also something else, that gives what we have experienced a deeper perspective. The atmosphere round the deathbed was not a passing over, as is often the case when a person is going to die, harassed, artificial. We found it quite the reverse, so natural and nearly to be taken for granted of, in all its dignity. It was precisely how Ole Krogness should die without any fear, marvellously confident, and victorious. His death crowned his life, as we knew him in his home, among friends, as fellow worker. Simple, sincere, pious, a central figure in all problems and practical duties that otherwise took up his time. An immediate kindness and warmth, and his assuring confidence every time we went to him for help. For me personally he meant a very great deal both in friendship, and as a fellow worker. On the one hand, his spiritual far-seeing superiority, which had no boundaries, and on the other hand his totally positive position in life, which led to practical self-control. What a consolation his calmness always was, "everything is so simple". It is goodness that has something to do with it. I always felt as if that man had his roots in life's perpetual source of energy. There was something about him that made one think he lived in two worlds. Strong bonds bound him to the earth, and a meaningful vocation went with him, both into the home and into society. But his longing went further, his home was also part of this world, where values are eternal, not just his music, which to him was a temple's forecourt, but was also a religious temple sanctuary.

And it was just that which became consummated on his deathbed. Engaged right until the end with everything that had filled his life, his close family, friends and work; how hard to give that up. And at the same time to behold the everlasting spirits' world, and did it not shine in his eyes of purity and peace that which shone out to meet him. It is wonderful to live but even more wonderful to die, he said,

at the door to an everlasting world. His spirit had reached there, where the boundary between life and death is no more.
Therefore it is not, in spite of everything, as if we have lost him. He is dead, but he is not far away from us. It is not just remembrances of him that live, but also himself, that spirit which was strong enough to meet death's horrors with a smile, will go with us through life.

We can feel this now, as we are assembled here after his funeral. He wanted us to be festive, not grieve. But mid in sorrow we feel something of that consolation and joy that lies in our awareness that a person has won a victory. We drink to his memory and note that it is as if he greets us from the other side.

Dagny and Olaf Devik were life-long friends of the Krogness family and often invited them on a visit. Archive: Private: Mette Bunting

Olaf Devik was a tremendous help for Dagny and the children after Ole' death. Olaf gave his support and help, and in connection with Ole's wishes he formed a group with Helland-Hansen, Rev. Irgens and Rev. Skagen to make sure that the family's affairs were put into effect after his death.

Nevertheless, there were changes. Previously Dagny felt safe in the knowledge that her sons would be educated, but now she knew that her widow's pension would not cover this. In the years ahead the Devik family gave Dagny and her children invaluable support. They often invited them to join their family during the summer holidays, and looked after them throughout all the years.[550]

In her memoirs for her grandchildren Dagny writes this about her everyday life during the first period after Ole's death.

> Oa had taken the General Certificate A-levels and moved to Oslo to the "*Studenthjemmet*", and received a grant for boarding costs while he studied, and the room cost 20 kroner a month, which aunt Julie paid, she was his godmother. Oa was a blood donor for ten years, and had pocket money from this. I had 5000 kroner to live on, and the children got free places at school, and so we managed. We knitted Fana-jackets, which we sold for 20 kroner each for the work. One year we had school children as boarders and that helped towards cost of food, as we had to have proper food for them. But when I was alone I had to give the maid notice, as I could not afford to have such help. The children were exceptionally good at helping me, and after taking their School Certificate, they found jobs and paid for living at home. Dagny and Olaf Devik who were also in Bergen were a great help to us, and a lot of pleasure in those difficult years.[551]

Dagny continued to live in Bergen until the beginning of World War II when she moved to Oslo. She lived there until she died in 1978, 89 years of age. She talked about her life on Haldde and in Tromsø until she died, as being the best years of her life.

---

[550] Dagny and Ole's daughter Eva told about this.

[551] Memoirs for grandchildren, written by Dagny, 1976.

# CHAPTER TWENTY SIX

# SUMMING UP

In this snow-covered building a great variety of joint activities took place, which gave far-reaching results. Photo Hilding Köhler. Archive; The World Heritage Rock-Art Centre, Alta Museum

*[...] everyday life is both the most significant and least esteemed part of people's history. (Troels-Lund 1914)*

In this final chapter we will look away from empirical descriptions, and consider what we have done, using analytic concepts from the field of research that we have explored.

We chose everyday life as a firm anchor point to convey important events in North Norwegian history, concerning the life of women and Northern Light's research on the summit of a mountain. In a way the choice was already made for us when we were given access to a pile of private letters written during the period 1912–1919, the majority of them concerned with a woman's experiences. The letters were full of everyday events, running parallel with research activities. The mountain top was important, but there was so much more. We have been engaged with the activities surrounding the world's first Northern Lights' observatory, and an everyday life, which was in retrospect praised by those who lived there.

When we delved into the written material the story unfolded before us and we saw issues that stretched further out into society as a whole. The topics involved in everyday life gave us the opportunity to look at larger contexts. The story developed. Activities on the mountain led to the establishing of a geophysical institute and weather forecasting services for Northern Norway, and the beginning of a university in the region. We wanted to find out about these people who made such a contribution to the academic development in Northern Norway. We were especially interested in shedding light on women's contributions. We wanted to contribute by filling the empty room in the existing literature about the activities on Haldde.

In this investigation we have kept closely to everyday life, care and work situations surrounding the home. In short, a life-world where our central characters communicated with each other and the world around them. The home represents the intimate, private, comprehensive and personal in society in contrast to the bureaucratic, objective, effective and specialized.[552] Nevertheless, people also act within a world system that consists of society's political, cultural and economical structures. The concept system-world indicates a system where money and power is administrated, and actions are executed to realise a specific aim.[553] In our material we find that a system-world filtered into everyday life in a way that produced rather long-drawn out results. One example about how the macro level in society was to be found in what took place at the micro level on Haldde can be seen in how the rooms in the house were used, and were described in national-wide newspapers and read about in Kristiania. Admittedly it was the researchers themselves who had written about this.

Life-world is the world we live in and carry out everyday activities, the normal life that each of us lives, not just at home but also at work and in society. It is an inter-subjective world where people first and foremost have a practical interest.[554] An everyday perspective emphasizes the action aspect of existence, and what different actions mean for the individual.

In a portrayal of the everyday we can find both women's and men's traditional and complementary work of that time. We also find power

[552] Gullstrand 1989.
[553] Habermas 1981.
[554] Schutz 1975.

differences and power hierarchies. This era was a period when conditions of power were changing through the dissolution of the union with Sweden, the development of Norway as an industrial nation, the growth of the labour movement and the women's movement. A milestone for women was reached in 1913 when they were given the right to vote. We wonder why Dagny does not make any comment about this in her letters to her mother-in-law as she usually writes quite frankly about things. Was she more concerned with this before she came to Haldde? Or was there just not room for such information in her daily struggles with her work? We found an interesting interpretation that explained women's reticence, which was formulated by Birkeland's close friend Kaja Geelmuyden in an article she published in an English women's journal. The example showed how a subtle power situation between women and men was found even within a common political affiliation.

This was a period of strong optimism for the future. New technological discoveries made it possible to "tame" nature in new ways, not least through large-scale development of hydropower, which took place all over the country.

Photo: Ole Andreas Krogness. Archive: National Library of Norway - Photo collection.

That Birkeland established a Northern Lights' observatory on the summit of a mountain in Finnmark, and initiated the research there of the Northern Lights, was also an expression for the optimistic spirit that prevailed at that time.[555] Optimism and belief that larger projects could be carried out and succeed, seems to persist in the Haldde colony despite a world war breaking out.

We took the Haldde colony's everyday practice as our starting point, their mastering of extreme living and weather conditions, and how their existence developed while they worked towards specific aims. Creativity and artistic development was in no way missing among the personalities on the mountaintop. One can imagine that music was a source of mastering, which made their life lighter in the middle of the loneliness and isolation. Dagny writes that music frightens away everything that is dark and sad.

Obviously, it was important for Ole that Dagny was satisfied with her everyday life, and had enough help and someone to talk to. That was probably also a primary reason for arranging for Olaf Devik and his wife Dagny to move up to Haldde. Socially it became more lively both when they arrived, as well as when the Lukkassen family with their three daughters moved into the house. It is most likely that thoughts about their own welfare were kept in mind, in connection with continuing their activities in Tromsø.

We have specifically been occupied with the supra-individual's cultural influence on the social activities of our main subjects, in accordance with the theoretical background perspective we initially sketched, at the beginning of the book. In what way are cultural differences expressed in social classes within a framework of everyday life on a mountain summit, and later in Tromsø? In this case it is divided quite harshly between education and non-education in the letters, when applied to assistants and maidservants. Dagny actually uses the term, "half-educated", without clearly showing what this exactly means. She uses the expression, "she has better parents", about a maidservant. It is shown in the material that like the rest of society at that time, the family one belonged to, and the education one had, were valued highly and sought-after. In small drips we find statements, which confirm a somewhat strained distance to local culture. There were also class distinctions within the research colony. The letters contain descriptions of social gatherings where everyone is not equally involved. Associating with the people living at the bottom of the mountain is limited to the local upper middle classes; the shopkeeper,

[555] Eikeset 1998:13.

clergyman, doctor and teachers,[556] but first and foremost to those who could play an instrument.

Connections with their own family, and professional contacts in the Norwegian scientific milieu, were actively maintained. The researchers also kept contact with foreign academic communities in the same field of study. This consisted of scientific research within the areas of Northern Lights, terrestrial magnetism and weather conditions. It was interesting to find out how close that network was and how well informed the network participants were, despite slow postage, shortly composed telegrams and long travelling distances. Ole's mother was important for the exchange of information. Regular contact with the governmental departments and other public institutions was important for financing their research, but just as important was the connection with wealthy private people.

They had a task on Haldde and this was to be carried out. Even so this task gave room for new creative ideas and clever solutions to major challenges. There were constant activities within the social network of individual people and the scientific milieu.

We have tried to interact between the detailed micro processes, which everyday life offers and which are experienced subjectively, and the more comprehensive structural processes, which were the framework conditions outlined for each individual's day-to-day existence. In an anthropologist perspective we find that everyday life is referred to in this way: "Even when the aim is to identify supra-individual cultural guidelines of actions, these should be linked to the individual's statements, behaviour and interaction in specific definable contexts."[557]

In our material, such supra-individual cultural guidelines are especially stated in the power relationship between housewives and maidservants. Both Dagny's actions and comments about the maidservants were rooted in an upper-class culture, which she handled to the best of her ability. But the hierarchy system internally in the women's milieu was in a process of change at the beginning of the 20th century, something which Dagny experiences through some of the maids' protests. Unfortunately, we do not have any written first-hand sources about how the maids on Haldde experienced the situation themselves. Neither do we have any first-hand written experiences about either the research assistants or their wives. We have concentrated on the letters that exist or used oral sources, archive material and literature. Quite a lot of the written sources in Northern Norway were burnt during World War II. The maids letters and diaries as well as what was written by others from Haldde have hardly been

[556] In accordance with an oral source: Werner Daniloff, Kåfjord.
[557] Gullestad 1989:145.

preserved as well as those written by researchers, and Dagny's own letters. We have made use of all available sources.

Kristian Birkeland, "*The Founding Father*", was given a central role in our work, as we gradually realized from the written material how much he meant for the researchers on Haldde. They greatly appreciated him, and showed their solidarity through phases of criticism and controversy. He had introduced Ole and Oaf to the world of research, and given them a professional task within the natural science academy. They were very dependent on his support, which they also received. While at the same time he simultaneously moved on with his own scientific work, and his ingenious ability to discover and exploit side effects of research experiments. Birkeland was an important trendsetter for the scientific milieu in the northern regions of Norway. In the sources there is a remarkable silence about the circumstances surrounding his death. The interpretations we make are based on material that goes far beyond the written communication between Haldde Mountain, and the surrounding world. We can assume that cultural values, and the development of middle-class ideals of that time contained taboo areas that determined what one should keep quiet about.

From published material about Birkeland we find a number of traces of contemporary cultural trends. Among other things we find traces that point towards a tension between the rationality in scientific knowledge and popular superstition. It was a surprise to find documentation about spiritualism, and that it was also used among people within a scientific milieu as late as the 20$^{th}$ century.[558] Even more interesting was finding out that the well-known suffragette, Fernanda Nissen, used spiritualism to come in contact with her beloved Oscar after his death in 1911.[559]

Our presentation is mainly connected to specific people, places, time, actions and events. That is in compliance with traditional anthropological documentation.[560] Even so we move beyond this and attempt to join together the information from different, and not necessarily comparable sources, into an overall correlation between the everyday life of individuals and over-individual cultural structures. We support our interpretations using specific professional literature and source material.

In the presentation of what happened during out grandparents' time, we are unavoidably wearing grandchildren's "glasses". We have different viewpoints, both in connection with geographically belonging, childhood and family connections. We have tried to be aware of the interpretations we

[558] Egeland & Burke 2005.
[559] Iversen 2009.
[560] Gullestad 1998:149.

take with us into our research. As interpreters of empirical material we will always find ourselves between the empirical events and the story we tell.[561] It is a position that requires reflexive attention, while at the same time being accountable to our source material.

Important questions in everyday life research are:

1. How does one create, maintain and convert background conditions?
2. How do people relate actively to different background conditions?[562]

We have pointed out what background conditions surrounded the Haldde activities, and tried to find out something of the historical development that created these conditions. Looking at the whole perspective, we find it was a coincidence that began all of this, starting with a young woman finding copper ore on a mountain above Kåfjord. That was the prelude, both to the Copper Works and to Northern Lights' research in the village and on the mountain, which became the birthplace for geophysical research. Even the expansion of hydropower and the larger enterprise, "*Norsk Hydro*", can be traced back to Kåfjord, with its waterfalls and hydroelectric construction, which was in lack of a power switch. We have also tried to show how the people we tell about, contributed each in their own way to both maintain and reshape the framework conditions, through practical and scientific work in the everyday lives they lived. Their life-world and the letters they wrote were contemporarily imprinted in cultural and political assumptions.

Here is an example of maintenance and conversion of background conditions: A maidservant on Haldde is dissatisfied with her work conditions. She is very young and her mother complains to Ole in a telephone conversation, using rather strong language, because her daughter does more work than just looking after the children, which she thinks is her actual job. The girl's mother demands that the girl should be sent home if the work situation does not change. Ole was caught off guard, but he attributes the bad language to being that, "they use strong language generally here in the north". He advises Dagny to keep her composure, and to calm down the situation, by talking to the girl. Find out what is bothering her and give her a raise in her wages, so that she is happy and satisfied. By transforming the crisis situation in a way that is acceptable for both sides, he contributes to avoiding an escalation of the conflict and a bad reputation. No other than the two of them, who meet socially in a heated telephone discussion, will be hurt.

[561] Jenssen 2009:19.
[562] Gullestad 1989:170.

The telephone conversation can be called a clash between the social system's organizational principles, employer and employee. Rather than maintaining unacceptable work conditions he chooses to agree with the mother and daughter. In that way a social structure, which is the frame for the power relationship between housewife and maidservant, can be won by being pliable; then only the angry mother has lost her composure. The framework conditions soften, the social structures are protected, and the power relationship is maintained.[563] Daily life continues as before.

A similar strategy cannot necessarily be used in all cases. Another maidservant left her employment on Haldde when her wage was not increased. People act differently to the framework conditions, which regulate their work. It is necessary to understand the social actors as active subjects with their own evaluations and choices.

It is especially in ethno-methodology that one treats people as acting subjects. The ethno-methodological direction arose in a sociological environment on the North American west coast as a reaction against structural functionalism, positivism and modernity theory, at the beginning of the 1960's. An ironic comment to the established academic community was that for the positivistic sociologist, "the subject is not much more than a zombie-like creature sleeping until it is woken by researchers with their questionnaires, and asked to mumble single word answers to a series of questions about certain aspects of social life".[564] Together with other creative sociologists ethno-methodology became a foundation for many later studies of everyday life, with a focus on the experiences of the subjects.

In the North American everyday life research there arose the term "sociological phantasy".[565] The concept shows that it is not just data collection or abstract theory that is of use when a problem in research is to be formulated. Fertile elements in a research process are among other things, personal feelings, restlessness and amazement together with aesthetic experiences.[566] In our project we have sometimes let a sociological imagination flourish, while simultaneously keeping firmly to the material. Many things point out that the researchers on Haldde also benefitted from conditions that lay outside their data collection, and abstract theories, in their formulation of research questions within their field of interest.

---

[563] Kristiansen 2005:213.
[564] Lyman 2002:15, quoted in, Hviid Jacobsen &Kristiansen 2005:255.
[565] Mills 1959/2002
[566] Gimmler 2005:96

We have made a point out of research and daily life being under the same roof, except for the observations and measuring, which was done from the actual observatory, even higher up on the mountain. The closeness between the two workplaces, housekeeping and scientific work also led to a female and a male dominated work arena being brought closer together. At the same time it entailed that society's micro and macro structures were mixed, and that men and women constituted a community. Whether the women's work influenced the actual research is not known. Nevertheless, Ole was obviously updated about how one helps a child into the world, when the local midwife did not come in time for the birth. In many cases there were both practical and enjoyable situations from living in the same locality. According to the letters it appears that the first period in Tromsø was difficult for Dagny, among other things because Ole 's place of work was a long way from their home.

A central unit of analysis in the everyday study, within this tradition, is the focused social meeting, which is the situation where people cooperate with each other face to face, and maintain a common focus of attention. Here situation definitions are expressed, negotiated and maintained. Through actions and gestures we also propose unavoidable suggestions as to how others perceive and treat us. We engage in "impression management".[567] When women's and men's work spheres are mixed, as on Haldde, such processes gave a common arena, which quite possibly created good mutual understanding of challenges each of them met.

Also publically, "impression management" was important. When Ole was preparing to go to Paris to give a speech about the case for "terrestrial magnetism", he felt that he ought to go out and "equip himself" so that he was distinguishably dressed for the meeting with the Parisian conference participants. He obviously wanted to present himself in a particular way. A central analytic perspective is about how everyday life contains role-play and dramatic performances.[568] This perspective can shed light on which mastering strategies were put into use by the Haldde researchers to succeed with their plans and ideas. In any meeting, a person's façade will appear. Such a façade consists of attitudes, appearance and "adornment", that we consciously or unconsciously use to present a specific image of who we are.[569]

Human encounters and social situations can be both a fatal and dangerous "game". It can result both in winning and losing. What we find in the letter material and in other documents, suggests that the Haldde

[567] Goffman 1959, ref. in Hviid Kristiansen 2005:199.

[568] Goffman1959

[569] Fine & Manning 2003:46, ref. in Hviid Jacobsen& Kristiansen 2005:198

researchers mastered this, and achieved good response to what they presented. With great gentleness they delt with scientific controversies and difficult situations. They managed to implement what they had set out to do. Perhaps due to their mastering of impression management and being able to remain cool, they also got as much as possible out of social meetings. This can have been a significant factor for their success. As Olaf wrote in his book, "we never complained!" A humorous, but serious basic tone may represent the crux of the matter for their mastering strategy.

It is worth mentioning Dagny's comment in her letter to her mother-in-law in 1913 that she "had her plans", in connection with the intended extension of the house on Haldde. She probably had a strong influence when it came to the realization of the new plans, which entailed getting more residents to live there. Her "impression management", with regard to the maidservants is however not so successful. An interpretation of this could be that the housewife's role collided with the beginning of the maidservant's rebellion at that time, which provided a more political undertone, of which Dagny seems unaware.

In everyday life studies, one can identify how people, often under stress, and in "fatal situations", have the character or ability to show self-control and dignity.[570] When Dagny, during the birth on the mountain grips the lamp and holds it, to "be of help" for Ole and her mother so they have a better light, while she pushes out the child, shows a reasonable level of self-control and dignity, in an extremely distressful situation. Mrs Devik also showed uncommon self-control, when she gave birth quietly, and alone in the bedroom, while the midwife sat outside and drank coffee.

One can say the same about Ole, who on his deathbed called all his children together and sang with them, and gave them encouragement about their new family life after he had gone. Likewise, when he talked to professionals and partners, and instructed them about how they should continue to work after his death, and claims with a smile that, "it is wonderful to live, but even more wonderful to die". From where did he get such self-control and dignity? We know from one of his last articles that he was concerned about the divine. However, it is the relationship between religion and science that occupies him. He writes that the admirable respect for the divine creation among, "old people, was so strong that it was regarded as a sacrilege to reason rationally about the heavenly phenomena".[571] He then goes through the theory about the universe's development and concludes with:

---

[570] Kristiansen 2005:206.

[571] Krogness 1932:321.

> Since the turn of the last century there has been a rapid development in all fields, the one significant discovery has followed after the other. Our conception of nature and its law has been expanded and transformed in many ways, phenomena one earlier had no knowledge of has been shown to play a role in many fields, in a way that no one could have dreamed of before.[572]

Here Ole sums up, what he himself has taken part in, which is what our book is mainly about. Ole's article starts with thoughts that spring out of everyday experiences, for example when one "on a clear night looks up at the twinkling heavens with its wonderful starry splendour."

Before we round off, we will take a look at the project in the light of its historical perspective. Knut Kjeldstadli's historical presentation is used as the basis for this. History as a subject is about human society, which spans over the actual lives of those who lived there in the past; how they behaved and the relationships they had to each other. It is about understanding people, interpreting what they do and say and grasping the reason for their activities. It is about connecting together what is told, into a greater overall picture, and regarding people as representatives of a larger environment, so that their lifecycle becomes the focus and mirror of the society that surrounds them.[573]

Our main material is about a wife living in a research colony on Haldde, and the relationships between all of those that live there. This has been placed in a local social context using the village of Kåfjord and the settlement in Bossekop as a background. Further, we have studied the contents of our material as a part of the national project at the beginning of the 20th century, when the established scientific institutions existed parallel with the political and the feminist movements. Additionally, we have studied how the activities in our research material has been affected by world events, such as World War I. Throughout the text we have attempted to give a portrayal of how the concepts 'structure' and 'actor' are combined socially and culturally in our material. We have also reflected on what is left unspoken in the letters.

In this way our text complies with newer historical knowledge, where events and people must be included, if one is to understand how new structures are created. This coincides with a general social and political interest for a more subjective approach and for culture and social anthropology, which we have touched upon earlier. Such a reorientation of

[572] Krogness 1932:134

[573] Kjeldstadli 2007:30.

history as a subject has opened up new ways of using oral sources.[574] We have interviewed people that have told us about their memories or what they knew about events in the actual region and in the historical period we have been concerned with. One interview about childhood memories from Haldde we have evaluated as being so special that we have honoured it with its own chapter, even though it touches upon the other themes in the book.

However, our main source is Dagny, as she presents herself in her letters, diary and memoirs for her grandchildren. Letter quotations have become the nerve centre in, and basis for all the chapters in the book. Memories and personal experiences in the memoirs as sources are, quite naturally always influenced by a person's frustrations, annoyances and other emotional reactions that have been left behind a long time ago. [575] While in Dagny's letters she looks towards the future, her memoirs are built upon her experiences viewed from a retrospective perspective. This gives a dynamic duality in the narrative presentation that supports each other, and which we have benefitted from in the book. Ole's letters are also first-hand sources. The archive material and documents have been used as complementary information for the main material. Several sources are independent of each other, and therefore support the credibility of the material.

Throughout the everyday theme we have shown the connection to the growth of new scientific institutions, such as the Geophysical Institute in Tromsø, and the Weather Forecasting Service for Northern Norway. Both were corner stones in the development of a University in Tromsø, which today is entitled UIT- Norway's Arctic University, and has merged with Finnmark University College in Alta.

What began with newcomers arriving in the village of Kåfjord at the beginning of the last century, where a "Science Company" had long been established to study the Northern Lights, resulted in a faculty under the university that was finally established in Tromsø in 1972. Based on our background material, we can see that for a while there was a question of where the geophysical research activities should be continued after the Haldde period; either in Alta or in Tromsø. What decided this is unclear. A significant factor could have been Dagny's wish for a larger social environment and Ole's wish of having a satisfied wife.

We have kept to human traces. The letters communicate deliberate messages; some are clearly normative, where Dagny expresses frustration and anger over the maid's incapability. Some letters are symbolic or

[574] Kjeldstadli 2007:85.
[575] Garro & Mattingly 2000.

illustrative, for example where she tells about practical tasks or gives beautiful descriptions of nature. Some letters are forward-looking, where the couple tell about their plans and what is to happen. Some evaluate what choice to make, for example to travel south or remain on Haldde, give birth on the mountain or in Bossekop, only write about positive things, or "quite naturally", about what they experience, both about sad moods and happy occasions. Certain letters are performative messages, for example Ole's application for the professorship at the University in Bergen. An invitation to selected citizens of Tromsø to a meeting to discuss the establishment of a Scientific Academy for Northern Norway is as far as we can see, also a performative notification, as too is the key speaker's thank-you speech given at the farewell dinner for Ole. These are at the same time actions where something is done.

Throughout the book we have used a large amount of quotations to be closer to our main informants. We wanted to create a place for the events, and let the reader have the opportunity to follow people in their everyday lives.

The main components in the material, as we see it now, are:

1. Ways towards the realization of scientific goals: networking, friendship, loyalty, humour and creativity, in addition to the hard work necessary to realize such goals.
2. Women's everyday work was an important prerequisite for research activities.
3. Changes in the relationship between housewives and maidservants.
4. Class divisions were changed by the Women's Movement and by different groups and individuals aiming for common goals, ordinary rights to vote, and better terms for all women.
5. A culturally decided silence around tabooed territories.

Through an everyday life perspective we believe we have shown something, which the sources themselves did not say anything about. This perspective gave us entry into how everyday actions could influence processes to which they were related, and give results higher up in the systems, while at the same time the systems' conditions ruled the activities into recognized tracks.

Work with this book has resulted in a greater total understanding of how the summit of a mountain in Kåfjord, through everyday activities stayed connected to the rest of the world. It is a constructed whole, built on the pieces we obtained from a range of sources that were at hand, about real but retractive events in history.

Thus, we have come to the end. The main questions for the project has been: What was it like to live on the top of Haldde Mountain? Why were these people there and where did they go after staying there for six years? We wanted to find out what happened and what was the result of the activities that took place 900 metres above the village by the fjord near Alta in Finnmark, Norway's most northern county. In this book we attempt to give answers to these questions.

Archive: National Library of Norway - Photo collection.

# OVERVIEW OF SOURCES


**Alta Museum**
Krogness, Dagny, diary
Krogness, Ole, letter to Birkeland, August 1915

**Newspaper Article**
Grønbech, D (2012) Gitta Jønssen – a woman pioneer.
Newspaper: Nordlys. Tromsø. 8th March, p.3

**Interviews carried out by authors**
Amtrup, Eva Asker 01.01. 2004
Dagenborg, Gudrun, Kåfjord, September 1994
Daniloff, Werner, Kåfjord 14.07. 2008
Holbæk-Hansen, Synnøve, Bergen November 1995
Lukkassen, Mary and Borghild, Oslo, April 1991
Opgård, Åse Synnøve Alta 15.07. 2008
Pedersen, Rannveig, Kåfjord 17.07. 2008

**National Library, Handwritten collection, Oslo**
Kjeldsberg, Betzy unpublished memoirs
Birkeland, Kristian, letters from Egypt
Devik, Olaf, private letters to Bjerknes, 1915–1920
Krogness, Dagny letters from Haldde and Tromsø, 1912–1926
Krogness, Ole Andreas, letters 1913–1928
Magazine "Nylaende" digitized

**Newspaper cuttings, private property of the Krogness family**
Aftenposten 14.02.1914, "From Haldde" by Olaf Devik
Aftenposten 27.10.1999
Bergens Tidende 23.08.1837
Bergens Tidende 23.08.1837 *Geofysikerne hedrer professor Krogness minne*, (The Geophysical Institute honoured Professor Krogness' memory)
Bergens Tidende 26.09.1979. *Hvordan Jan Mayen ble en del av Norge.* (How Jan Mayen became a part of Norway)

Bragd number.16/1941. *Et idealsamfunn 1000 m. over havet, dagliglivet på et sylkvast finnmarksfjell (*An ideal society 1000 metres above sea-level, daily life on a sharp-peaked mountain in Finnmark)
Finnmark Dagblad 19. 11. 1983
Kristian Schjelderup, also reported in Nordlys 30.06.1934
Nordlys 17. 01.1981
Tidens Tegn 02.061926
Tidens Tegn 12. 2.1932
Tromsø 30.07.1928 (Uncertain about the name of the newspaper, because of missing information
Uniternt *(*unknown date*) Det var godt å bo på Hal'di!, møt norsk forskings grand old man: Olaf Devik.*(It was good to live on Hal'di!, with Norwegian researchers' grand old man: Olaf Devik)
Vårt Blad nr. 3 13.03 1991
Vårt Blad number. 3,13.03. 1991 *Nordlysslottet på Haldde* (The Northern Lights' Castle on Haldde)

- Different newspaper cuttings used from 1934
- Aftenposten 29.05.1934
- Aftenposten, Lars Vegard 30.05-1934
- Bergen Aftenblad 29-05-1934
- Dagbladet – unknown date
- Mathilde Mathisen, in an unknown newspaper
- Morgenbladet – unknown date
- Tidens Tegn, Carl Störmer 30.05.1934

**The National Archives of Norway, in Oslo**
Box 578
*Kirke- og utdanningsdepartementet i skolekontor* D, Eh 0581, Eh-0578, Eh-0579 (The Ministry of Church and Education – in the School Office)
Private archive 686, Olaf Devik, Numbers 11, 12 and 13

**Written sources belonging to the families**
*(Krogness, Guldberg and Devik)*
Devik, Dagmar, Dagny, Ingebjørg and Ole, *Samtale med farmor* (Conversation with grandmother) 27.08.1985
Devik, O. (1915–1920) *Private brev i familien Deviks eie (*Private letters belonging to the Devik family)
Devik, O. (1934) *Hendelsene rundt Ole Krogness's død* (Events surrounding Ole Andreas Krogness' death)
Devik and Birkeland Various letters from them

Diverse *dikt, sanger og gåter diktet på Haldde* (Different poems, songs and riddles written while on Haldde)
Evjen, Sigurd, *brev til (*letters to) Dagny Krogness
Guldberg, Carl Johan, *brev til Det Norske Missionsselskab* (letter to the Chairman of the Board of the Norwegian Missionary Society), 1876–1886
Committee *for De Vitenskapelige Institusjoner* (the Scientific Institutions) in Tromsø 1928–1953
Krogness, Dagny, *Erindringsnotat til barnebarna* (Memoirs for the grandchildren), 1976
Krogness, Ole Andreas, *postkort sendt fra reisen med Birkeland til Egypt* (postcard sent from Egypt when travelling with Birkeland)
Krogness, Ole Andreas, *søknad til professorstilling* (application for the Professor position) in Bergen
ODE *fra Værvarslinga i Nord-Norge* (from Weather Forecasting Service in Northern Norway)

**Statsarkivet, Tromsø (The Regional State Archive in Tromsø)**
*Haldde Nordlysobservatoriet, Mappe 7, Boks 40* (Northern Lights' observatory on Haldde File 7, Box 40)
Krogness, Ole, *årsrapporter* (Annual reports) 1913–1918
Værvarslingas *arkiv – Observatoriet på Haldde i Kåfjord-Kopibøker* (Weather Forecasting's Archive – Observations on Haldde in Kåfjord-Copy books)

**Universitetsbibloteket, UiT Norges arktiske universitet (University Library, UiT Norway's Arctic University)**
Birkeland's *rapport i Morgenbladet digitalisert utgave* (digitized newspaper report) 18.08.1912,

# BIBLIOGRAPHY

Alta Museum (1999). *Bruks- og restaureringsplan for nordlysobservatoriet på Haldde.*

Alvesson, M. & Skjöldberg, K. (1994). *Tolkning och reflektion. Vetenskapsfilosofi och kvalitativ metod.* Lund: Studentlitteratur.

Barlaup, A. (red.) (1966). *Det Norske Meteorologiske Institutt 1866–1966.* Oslo: Fabritius og Sønner.

Beichman, M. (1939). *Forunderlige Finnmark.* Oslo: Gyldendal.

Bloch-Hoell, N.E. (2009). Kristian Vilhelm Koren Schjelderup. *Store norske leksikon.* Hentet 19.08.2014 fra www.snl.no/Kristian_Vilhelm_Koren_Schjelderup

Blom, I. (1973). *Kampen om Erik Raudes land.* Oslo: Gyldendal.

Bore, R.R. (red.) (2007). *På liv og død, Helsestatistikk i 150 år.* Statistisk Sentralbyrå. Hentet 19.08.2014 fra http://www.ssb.no/a/publikasjoner/pdf/sa94/sa94.pdf

Borza, T. (2001). Spanskesyken i Norge 1918–19. *Tidsskrift for Den norske Legeforening.* Nr. 30, 121: 3551-4. Hentet 19.08.2014 fra http://tidsskriftet.no/article/462099

Bratteng, O. (1980). Radiobølgeforplanting i ionosfæren. *Ottar.* Tromsø: Universitetet i Tromsø, s. 38–61.

Brekke, A. (1982). Haldde – arnested for den moderne nordlysforskning. I *Altaboka. Årbok for Alta* (s. 56–75). Alta: Alta Historielag.

Brekke, A. & Egeland, A. (1979). *Nordlyset. Fra mytologi til romforskning.* Oslo: Grøndahl & Søn Forlag A.S.

Brekke, A. & Egeland, A. (1994). *Nordlyset kulturarv og vitenskap.* Oslo: Grøndahl og Dreyer.

Brekke, A. & Lynne, T. (1997). *Nordlys.* Alta: Alta Museums småskrifter, nr. 4.

Børdahl, P.E. & Hem, E. (2004). Endringer i norsk fødselshjelp i årene 1915–61. I *Tidsskrift for Den norske legeforening.* Nr. 24, 124: 3231-4.

Devik, O. (1915). Haldde-observatoriet og stormvarsler nordpaa. *Norsk fiskeritidende.* s. 200–203.

Devik, O. (1935). Professor O.A. Krogness og hans innsats i norsk geofysikk. *Naturen,* s. 161–167.

Devik, O. (1967). *Kristian Birkeland as I knew him*. Foredrag presentert på The Birkeland Symposium on Aurora and Magnetic Storms, Sandefjord.

Devik, O. (1971). *Blant fiskere, forskere og andre folk*. Oslo: H. Aschehoug & Co.

Devik, O. (1976). *Fra pionertiden i Norsk Fysikk og Geofysikk*. Tromsø: Universitetet i Tromsø.

Devold, H. (1940). *Polarliv*. Oslo: Gyldendal.

Drivenes, E. A. (1984). ... Hvert hammerslag nyt haab skal vinde... *Altaboka. Årbok for Alta* (s. 110–118). Alta: Alta historielag.

Drivenes, E. A. & Jølle, H.D. (red.) (2004). *Norsk Polarhistorie II, Vitenskapene.* Oslo: Gyldendal.

Egeland, A. (1994). *Kristian Birkeland, mennesket og forskeren*. Oslo: Norges Banks Seddeltrykkeri.

Egeland, A. (2012). *Nordlyspioneren Carl Størmer.* Fysisk Institutt, Oslo: Universitetet i Oslo.

Egeland, A. & Burke, W. J. (2005). *Kristian Birkeland, The First Space Scientist*. Dordrecht: Springer.

Eikeset, K. J. (1998). *Altas historie. Bind 3. Dramatiske tiår 1920–1964*. Alta: Alta kommune.

Elstad, I. (2006). I ei seng i stua. I I. Elstad & T. Hamran: *Sykdom Nord-Norge før 1940*. Bergen: Fagbokforlaget.

Elstad, I. & Hamran, T. (2006). *Sykdom i Nord-Norge før 1940.* Bergen: Fagbokforlaget.

Ericsson Ryste, M. & Wreden Kåss, I. (2013). *I mål*. Hentet 19.08.2014 fra http://www.stem-merett.no/imal/

Ertresvåg, V. E. (1979). Fyrstikkarbeiderstreiken. *Pax leksikon C-G*, Oslo: Pax forlag, s. 475–76.

Evjen, S. (1945). Professor O. A. Krogness in memoriam. I *Værvarslinga for Nord-Norge 25 år. Festskrift.* Tromsø: A/S Peder Norbye.

Fine, G.A. & Manning, P. (2003). Erving Goffman. I G. Ritzer (red.): *The Blackwell Companion to Major Contemporary Social Theorists* (s. 354–62). Oxford: Blackwell. *Forhandlinger ved de Skandinaviske naturforskeres trettende møde i Christiania 7. 12. juli 1886 (1887).* Christiania: Grøndahl og Søns Boktrykkeri.

Forsdahl, A. (2000). Finnmarkslegen som påviste arvens betydning ved hofteleddsdysplasi. *Tidsskrift for Den Norske Lægeforening* nr. 22. 120:2672-3. hentet 19.08.2014 fra http://tidsskriftet.no/ article/175963

Foss, A.S. (01.10.2003). *Tuberkulose på vei tilbake*. Forskning.no. Hentet 19.08.2014 fra http://www.forskning.no/artikler/2003/januar/1042022383.72

Friedman, R.M. (2004). Nansenismen. I E.A. Drivenes & H.D. Jølle (red.): *Norsk Polarhistorie II, Vitenskapene* (s. 107–174). Oslo: Gyldendal.

Fulsås, N. (1993). *Universitetet i Tromsø 25 år.* Tromsø: Universitetet i Tromsø.

Føyn, N. J. (1915). Utvidelse av observatoriet paa Haldde. – Stormvarsler i det Nordlige Norge. *Fiskeritidende.* s. 147–150.

Garro, L. C & Mattingly, C. (2000). Narrative as construct and construction. I L.C. Mattingly & L. Garro (red.): *Narrative and the cultural construction of illness and healing* (s. 1–49). Berkeley: University of California Press.

Geelmuyden, K. (1916). Women voters in Norway. *Votes for women*, Jan 28., s. 237. Hentet 18.08.2014 fra http://news.google.com/newspapers?id=ZW89AAAAIBAJ&sjid=Wys - MAAAAIBAJ&pg=6128%2C18963314

Giddens, A. (1993). *New sociological Method. A Positive Critique of Interpretative Sociologies.* Cambridge: Polity Press.

Gimmler, A. (2005). Amerikansk pragmatisme – Hverdagslivets kreativitet. I M. Hviid Jacobsen & S. Kristiansen (red.): *Hverdagslivet. Sociologier om det Upåaktede* (s. 72–105). København: Hans Reitzels Forlag.

Goffman, E. (1959). *The Presentation of Self in Everyday Life*. London: Penguin Books. Greve, M. (1906). *Mor og barn.* Kristiania: Aschehoug & Co.

Gullestad, M. (1989). *Kultur og hverdagsliv*. Oslo. Universitetsforlaget.

Habermas, J. (1981). *Theorie des kommunikativen Handel* 1–2. Frankfurt: Suhr-kamp.

Hansen, F. R. (1999). *Nordenfjelske 1857–1985 Fartøyene og deres historie.* Trondheim: Nordenfjeldske Dampskibsselskab Trondheim.

Hareide, J. (2009). Hanna Winsnes. *Norsk biografisk leksikon.* Hentet 19.08.2014 fra http://snl.no/.nbl_biografi/Hanna_Winsnes/utdypning

Heitmann, F. (2004). Y-13 og dens familie. I *Altaboka* (s. 115–122). Alta: Alta Historielag.

Heitmann, F. (ukjent år). Lauritz Heitmann Handelsmannen – kirkemannen – forretningsmannen. Hentet 19.08.2014 fra Sulithjelma historielag nettbibliotek http://www.sulisavisa.no/historielag/Bibliotk/index.htm

Helskog, K. (1988). Helleristningene i Alta i et fennoskandisk perspektiv, kronologi og symbo- lisme. *Iskos*. 7 s. 67–76.

Hviid Jacobsen, M. & Kristiansen, S. (2005). *Hverdagslivet. Sociologier om det upåaktede*. København: Hans Reitzels Forlag.

Hætta, O. M. (1992). *Samene – en arktisk urbefolkning*. Karasjok: Davvi Girji o.s.

Høegh Schmidt, H. (1912). *Veiledningsbok for tjenestejenter og unge husmødre.* Kristiania: Grøndahl & Søns Forlag.

Hølaas, O. (1945). *Norge under Haakon VII.* Oslo: J.W. Cappelen Forlag.

Iversen, I. (2009). Fernanda Nissen: utdypning. I *Norsk biografisk leksikon*. Hentet 19.08.2014 fra http://nbl.snl.no/Fernanda_Nissen

Jago, L. (2001). *Nordlysets gåte. Beretningen om Kristian Birkeland. Pioneren som ofret alt i viten- skapens tjeneste.* Oslo: Gyldendal.

Jenssen, T. (2009). *Behind the eye. Reflexive methods in culture studies, ethnographic film, a visu- al media*. Århus: Intervention Press.

Kirkehusmo, A. (2013). Første verdenskrig, *Store Norske Leksikon.* Hentet 19.08.2014 fra http://snl.no/Første_verdenskrig#menuitem6

Kjeldstadli, K. (2007). *Fortida er ikke hva den en gang var.* Oslo: Universitetsforlaget. Kjeldstadli, K. (2010). *Mine fire besteforeldre. En familiefortelling om framveksten av det moderne Norge.* Oslo: Pax forlag A/S.

Klausen, A. M. (1992). *Kultur – mønster og kaos*. Oslo: Ad Notam, Gyldendal.

Knutsen, N. M. & Posti, P. (2002). *La Recherche, en ekspedisjon mot nord*. Tromsø: Angelica.

Kristiansen, S. (2005). Ervin Goffman – Hverdagslivets rollespill. I M. Hviid Jacobsen & S. Kristiansen (red.): *Hverdagslivet. Sociologier om det Upåaktede* (s.194–221). København: Hans Reizels Forlag.

Krogness, O. (1914). Haldde-observatoriet, dets virksomhet og nogen foreløpige Resultater. *Naturen,* juli–august, 193–216, 303– 318.

Krogness, O. (1915). Svar til hr. Føyn i anledning hans artikel om Haldde Observatoriet i Norsk Fiskeritidende's maihefte. *Norsk Fiskeritidende*, s. 203–209.

Krogness, O. (1925). Seilas i tåke. Særtrykk av *Norsk Tidsskrift for Sjøvesen*, *43*, s.393–405 og 431–448

Krogness, O. (1932). Universets utvikling, *Naturen*, s. 321–367.

Kronen, T. (2009). Ellen Gleditsch. I Norsk Biografisk Leksikon. Hentet 19.08.2014 fra http://nbl.snl.no/Ellen_Gleditsch

Kübler, M. (1885). *Husmoderen i Staden og på Landet*. København: Nicolai Cohens Forlag.

Larsen, R. & Berger, S. (2009). Nordlysobservatoriet – historie og erindringer. Universitetet i Tromsø, Det Matematisk Naturvitenskapelige Fakultetet. Hentet 19.08.2014 fra http://munin.uit.no/bitstream/handle/10037/4392/book.pdf?sequence=1

Lindeman, T. (1924). *Den Trønderske familie Lindeman, samlet og skrevet for Familien*. Trondheim: Waldemar Janssen A/S Boktrykkeri.

Lyman, S. M. (2002). Restoring the Self as Subject: Addressing the Question of Race. I J. A. Kotarba & J. M. Johnsen (red.): *Postmodern Existential Sociology* (s. 15–41). Walnut Creek: Alta Mira.

M. Hj. (1917). Samerne i Norge og deres første landsmøte. *Nylænde* (digitalt), Oslo: Nasjonalbiblioteket.

Maffesoli, M. (1993). The Social Ambiance. *Current Sociology*. 41 (2) s. 7–15.

Mamelund, D. E. (2005). Spanskesyken rammet sosialt skjevt. *Samfunnsspeilet*. 2005/1, Statistisk Sentralbyrå. Hentet 19.08.2014 fra http://www.ssb.no/helse/artikler-og-publikasjoner/span-skesyken-rammet-sosialt-skjevt

Mills, C.W. (1959/2002). *Den sosiologiske fantasi.* København: Hans Reitzels Forlag.

Minde, H. (1983). *Stein og brød. Skifernæringa i Alta fram til 1940.* Alta: Alta Skiferbrudd, Alta historielag og Alta kommune

Moe, A. (2005). Russlands historie. *Store Norske Leksikon*. Hentet 19.08.2014 fra http://snl.no/Russlands_historie#menuitem5

Mykland, K. (red.) (1978 a). *Norges Historie, Norge i støpeskjeen 1884–1920*. Bind 12. Oslo: Cappelens forlag.

Mykland, K. (red.) (1978 b). *Norges Historie, Klassekamp og fellesskap1920–1945*. Bind 13. Oslo: Cappelens forlag.

Nasjonalforeningen for folkehelsen (2013). *Nasjonalforeningens historie*. Hentet 19.08. 2014 fra http://www.nasjonalforeningen.no/no/Om_oss/Historie/

Nielssen, A. R. (1984). Den norske kolonisering av Altafjorden på 1600-tallet. I *Altaboka. Årbok for Alta* (s. 100–109). Alta: Alta Historielag.

Nielsen, J. P. (1995). *Altas historie, Bind 2. Det arktiske Italia 1826–1920*. Alta: Alta kommune. Nielsen, R. (2002). *Gryr av dag. Jubileumsskrift for Troms Arbeiderparti 1902–2002.* Tromsø: Lundblad Media.

Nielssen, A. R. & Petterson, A. (1993). *Nordlyspionerene. Menneskene og observatoriet på Halddetoppen i Alta.* Oslo: Grøndahl & Dreyers Forlag

Nilssen, A. C., Brekke, A. & Ørbæk, J. B. (red.) (2006). *Arctic Lights, Our Natural and Cultural Heritage*. Tromsø: Tromsø University Museum.

Nissen, K. (1941). Recherche-ekspedisjonen. Et hundreårsminne. *Norsk Geografisk Tidsskrift – Norwegian Journal of Geography*. 8:5–6, s. 161–218.

Norsk Helseinformatikk (2013). *Kullosforgiftning,* hentet 19.08.2014 fra http://nhi.no/forside/forstehjelp/akuttmedisin/kullosforgiftning-29201.html

Paulgaard, G. (1997). Feltarbeid i egen kultur – innenfra, utenfra eller begge deler? I E. Fossåskaret, O. L. Fuglestad & T. H. Aase (red.): *Metodisk feltarbeid. Produksjon og tolkning av kvali- tative data* (s. 70–94) Oslo: Universitetsforlaget.

Pedersen, B. (2014). Guldberg og Waages lov. *Store Norske leksikon.* Hentet 19.08.2014 fra http://snl.no/Guldberg_og_Waages_lov

Pedersen, G. (2003). 100 år med redusert spedbarnsdødelighet, *Samfunnsspeilet. Statistisk Sentralbyrå.* Hentet 19.08.2014 fra http://www.ssb.no/helse/artikler-og-publikasjoner/100-aar-med-redusert-spedbarnsdodelighet

Pedersen, R.H. (2008). Marit i Dullan. I *Altaboka 2008. Fra fjell til fjære* (s. 54–62). Alta: Alta Historielag.

Repstad, P. (2009). Kristian Schjelderup. *Store norske leksikon.* Hentet 19.08.2014 fra http://nbl.snl.no/Kristian_Schjelderup#menuitem2

Rosaldo, R. (1989). *Culture and Truth. The Remaking of Social Analysis.* Boston: Beacon Press. Scheen, E. (1957). *«Kong» Thomas i Kåfjord.* Oslo: Trykt som manuskript.

Schutz, A. (1975). *Hverdagslivets sociologi.* København: Hans Reitzels Forlag.

Skard, T. (2013). Elsa Laula Renberg: Første politiske sameforkjemper. *Rødt Marxistisk Tidsskrift*, nr. 1. Hentet 19.08.2014 fra www.marxisme.no/index.php/2013/nr-1-2013/929-elsa

Sogner, S. & Telste, K. (2005). *Ut og søkje teneste – historia om tenestejentene.* Oslo: Det Norske Samlaget.

Statistisk Sentralbyrå (1961). *Dødeligheten og dens årsaker i Norge 1856–1955.* Samfunnsøkonomiske studier nr. 10. Oslo.

Statistisk Sentralbyrå (1966). *Dødeligheten blant spedbarn i Norge 1901–1963.* Samfunnsøkonomiske studier nr. 17. Oslo.

Statistisk Sentralbyrå (2000). *Statistisk ordbok.* Hentet 19.08.2014 fra http://www.ssb.no/a/hist-stat/aarbok/ht-0801-322.html

Statistisk Sentralbyrå (2001). *Mindre melk og brød, tallenes fortelling: forbrukervaner 1906 –1998.*

Hentet 19.08.2014 fra http://www.ssb.no/inntekt-og-forbruk/artikler-og-publikasjoner/min-dre-melk-og-brod

Statistisk Sentralbyrå (2005). *Statistisk årbok 2005.* Hentet 19.08.2014 fra http://www.ssb.no/a/histstat/aarbok/ht-0605-255.html

Statistisk Sentralbyrå (2013). *Konsumindeks.* Hentet 19.08.2014 fra http://www.ssb.no/kpi

Stenseth, B. (2000). *Eilert Sundt og det Norge han fant.* Oslo: Gyldendal.
Spanskesyken (2005). *Store Norske Leksikon.* Hentet 19.08.2014 fra http://snl.no/spanskesyken Rolad Amundsen (2005. *Store Norske leksikon.* Hentet 19.08.2014 fra http://snl.no/Roald_Amundsen Sæland, S. (1936). Kristian Birkeland. I *Mennesker jeg møtte: tolv radioforedrag.* Stryn: Bibliotekenes IT-senter AS.
Tallaug-Olsen, A. (1941). Nordlys-observasjoner og hjemmeliv på en fjelltopp 1000 meter over havet. *Urd,* 45 (3), [55]–56
Troels-Lund, T. F. (1914). *Dagligt Liv i Norden i det sekstende Aarhundrede,* Fjerde Utgave. Gyldendalske Boghandel. København: Nordisk Forlag.
Tromholt, S. (1885). *Under nordlysets straaler.* København: Gyldendalske boghandels Forlag.
Vammen, T. (1986). *Rent og urent. Hovedstadens piger og fruer 1880–1920.* Gyldendalske Boghandel. København: Nordisk Forlag.
Vegard, L. & Krogness, O. (1920). In Memory of our teacher Kr. Birkeland. I *«The position in space of the Aurora Polaris. From observations made at the Haldde-observatory 1913–14».* Geofysiske Publikationer. Vol. 1, NO. 1.
Velle, W. (2012). Turberkulose – veterinærvitenskap. *Store Norske Leksikon.* Hentet 19.08.2014 fra http://snl.no/tuberkulose/veterinærvitenskap
Vervarslinga for Nord-Norge 25 år (1945). *Festskrift.* Tromsø: AS Peder Norbye.
Wiig, F. L. (2007). Alt for et hygienisk Norge. Historisk helsestatistikk, Folkehelsearbeid 1910–1914.
I R. R. Bore (red.): *På liv og død, Helsestatistikk i 150 år* (s.153–163). Statistisk Sentralbyrå hentet 19.08. 2014 fra http://www.ssb.no/a/publikasjoner/pdf/sa94/del-v-2.pdf
Wikan, S. (2008). *Ellisif Wessel. En biografi.* Oslo: Pax Forlag.
Winsnes, H. (1845). *Lærebog i de forskjellige Grene af Husholdning,* Dokumentasjonsprosjektet. S. 153–163. Hentet 19.08. 2014 fra *http://www.dokpro.uio.no/litteratur/winsnes/frames.*htm
Ytreberg, N. A. (1962). *Tromsø bys historie. Annet bind.* Tromsø: AS Peder Nordbye.
Ytreberg, N. A. (1971). *Tromsø bys historie. Bind III.* Tromsø: AS Peder Nordbye.
Aakvaag, G. C. (2008). *Moderne sosiologisk teori.* Oslo: Abstrakt forlag AS.

Aas, H.M. (2011). *Organisering av jordmortjenesten – faglige idealer og politiske realiteter.* Masteroppgave. Hentet 19.08. 2014 fra https://www.duo.uio.no/bitstream/handle/123456789/30236/Master-H-M-as.pdf?-sequence=1

# EXPERIENCE

Splendid Northern Lights
Never seen them that beautiful
Cold, dancing flames
Capricious
Free rhythms
Parallel pointed barbs, tossed themselves
sharp
Bent down, Stepped forward
Joint together and spread out
Became weak, became strong
Became an arrangement, for a short second,
Posing
like a model
Change direction, form a new pattern
A movement artist
A Guest-visitor in the atmosphere
Then sudden, all is still
The stars alone hung there in their fixed positions
The wild and beautiful Northern Lights' were gone.

Toril – Experience September 2013
*(Translated from Norwegian by Toril Jenssen.)*

# PRIMARY AUTHOR FOR EACH CHAPTER

**Chapters:**

| | |
|---|---|
| Introduction | Toril Jenssen |
| 1. A Dramatic Entré | Mette Bunting |
| 2. Kåfjord's International History | Toril Jenssen |
| 3. Alliance with Local People | Mette Bunting |
| 4. Early Days on Haldde | Mette Bunting. |
| 5. The Enigmatic Genius | Toril Jenssen. |
| 6. A Dangerous Event | Mette Bunting |
| 7. Housekeeping at a High Altitude | Mette Bunting |
| 8. Professional Work and Daily Life | Toril Jenssen |
| 9. Dagny's Struggle | Mette Bunting |
| 10. Practical Hardship and Academic Controversies | Toril Jenssen |
| 11. The Main House Expands | Mette Bunting |
| 12. New Plans | Toril Jenssen |
| 13. A Child is Born | Mette Bunting |
| 14. Progress in the Research Colony | Toril Jenssen |
| 15. More Births | Mette Bunting |
| 16. Childhood Memories of Haldde | Toril Jenssen |
| 17. Social Life for Children and Adults | Mette Bunting |
| 18. Goal in Sight | Toril Jenssen |
| 19. Housewives and Maids | Toril Jenssen |
| 20. No Return | Toril Jenssen |
| 21. Facing Death | Mette Bunting |
| 22. From Mountain Top to City Life, and a Possible University? | Toril Jenssen |
| 23. Gender, Culture, Class | Toril Jenssen |
| 24. Further Work | Mette Bunting |
| 25. Life Moves On | Mette Bunting |
| 26 Summing up | Toril Jenssen |
| Poem: Experience | Toril Jenssen |